Cable Television:
Franchising Considerations

Rand Cable Television Series
Walter S. Baer, *Series Editor*

•

CABLE TELEVISION:
A Handbook for Decisionmaking
Walter S. Baer

CABLE TELEVISION:
Franchising Considerations
Walter S. Baer, Michael Botein,
Leland L. Johnson, Carl Pilnick,
Monroe E. Price, Robert K. Yin

CABLE TELEVISION:
A Guide to Federal Regulations
Steven R. Rivkin

CABLE TELEVISION:
Developing Community Services
Polly Carpenter-Huffman,
Richard C. Kletter,
Robert K. Yin

CABLE TELEVISION:
Franchising Considerations

Walter S. Baer

Michael Botein

Leland L. Johnson

Carl Pilnick

Monroe E. Price

Robert K. Yin

Crane, Russak & Company, Inc.
NEW YORK

Published in the United States by
Crane, Russak & Company, Inc.
347 Madison Avenue
New York, N.Y. 10017

Manufactured in the United States of America

601017

CONTENTS

Preface v

Chapter 1. A Guide to the Technology 1
            Carl Pilnick and Walter S. Baer

Section
  I. INTRODUCTION 3
      Cable Television Distribution Versus Broadcasting 4
      How a Basic Cable System Works 5
          Towers and Antennas 5
          Headend 7
          Facilities for Local Origination 7
          Cable Trunk 9
          Amplifiers 10
          Feeder Cable 11
          Subscriber Taps and Drops 11
      Extending the Range of Cable Services 13

 II. COMMUNICATION CAPACITY OF CABLE SYSTEMS 15
      The Basic 12-Channel System 15
      Expanded Channel Systems 17
          Multiple Cable Systems 17
          Systems with Converters 19
          Switched Systems 22
      The TV Receiver Problem 23
      Comparing Alternatives 23

III. CABLECASTING 25
      Studio Facilities 25
      Video Cameras and Tape Recorders 27
      Mobile Facilities 30
      Program Production 32

 IV. INTERCONNECTION OF CABLE SYSTEMS 33
      Forms of System Interconnection 33
          Internal Interconnection (Subdistricting) 33

Area Interconnection 33
Networking 34
Interconnection Techniques 35

V. TWO-WAY COMMUNICATIONS 37
Subscriber Services 37
Institutional Services 41
Two-Way Transmission Techniques 41
Two-Way Demonstration Projects 44

VI. TECHNOLOGY FOR NEW SERVICES 49
Pay TV and Private Channels 49
Information Retrieval and Frame Storage 52
Electronic Mail and Facsimile 53
Videocassette Recorders and Cable Television 54
High-Resolution TV 54
Institutional Services 55
Teleconferencing 55
Video Surveillance 56
High-Speed Data Transmission 56

VII. TECHNICAL STANDARDS 58
FCC Technical Standards 58
Writing and Enforcing Local Technical Standards 59
Signal Quality 60
Reliability and Maintainability 61
Performance Monitoring and Testing 61
Problem Areas 63

Appendix:
FCC TECHNICAL STANDARDS FOR CABLE SYSTEMS 65

Chapter 2. The Process of Franchising 69
Leland L. Johnson and Michael Botein

Section
I. INTRODUCTION 71
Objectives 71
A Note on Source Materials 72

II. DRAFTING AND AWARDING THE FRANCHISE 74
The Negotiation Approach versus the Competitive
Bid and Award Approach 74
Step 1: Adoption of Procedures for Drafting and
Awarding the Franchise 75
Step 2: Assessment of Community Needs, Objectives,
and Alternatives 76
Step 3: Hearings and Tentative Decisions Regarding
Major Issues 81
Step 4: Hearings on and Adoption of Draft Franchise 83

Step 5: Preparation and Dissemination of Request
for Proposals 84
Step 6: Hearings on Proposals 86
Step 7: Decision on Award of Franchise 87
Step 8: FCC Certificate of Compliance 90
Step 9: Monitoring System Construction and
Certifying Performance 91
Step 10: Continuing Administration of the Franchise 91

III. TERMS AND CONDITIONS OF THE FRANCHISE 93
Prefatory Provisions 93
Definition of Terms 94
Duration of Franchise 97
Geographic Exclusivity 99
Broadcast Signals to Be Carried 100
Construction Timetable 101
Extent of Wiring in the Franchise Area 102
Construction Requirements 104
Employment Practices and Training 107
Technical Standards 108
Operational Standards 109
Access to Premises by Cable Operator 111
Rates and Other Charges to the Subscriber: General
Considerations 112
Cable Connection Fees 114
Monthly Service Rates, Disconnection and Reconnection
Charges 116
Relocation Charges 119
Billing and Payment Procedures 119
Provision for Temporarily Reduced Charges 120
Reduced Rates for Special Classes of Users 120
Establishing and Adjusting the Rate Structure 122
Setting Rates and Connection Charges for Basic and
Ancillary Services 122
Review and Revision of Rates 122
Rates for New Services 124
Allocation of Channels 125
Pay Programming 128
Facilities for Public Access to System 129
Minimum Channel Capacity 130
Interconnection of Systems 131
Franchise Fees 132
Provisions for Transfer of Franchise 135
Performance Bonds 137
Liability for Damages by the Cable Operator 138
Reporting Requirements 139
Franchisor's Rights 140
Concentration of Control 140
Maintenance of Home Antennas 141
Emergency Use 142

    Separability of Clauses, Compliance with Applicable Laws 142
    Receivership 143
    Cancellation and Expiration 143

IV. CONCLUDING REMARKS 146

Appendix
A. CHECKLIST OF MAJOR ELEMENTS IN THE FRANCHISING
    PROCESS 147
B. SINGLE VERSUS MULTIPLE OWNERSHIP 151

Chapter 3. Technical Considerations in Franchising 157
          Carl Pilnick

Section
I.   INTRODUCTION 159

II. AREAS OF TECHNOLOGICAL UNCERTAINTY 163
     Background 163
     Technical Standards 165
     Channel Capacity 169
     Two-Way Communications 174
     System Quality and Reliability 177
     Interconnection and Networking 185
     Requirements for New Services 188

III. TECHNOLOGICAL UNCERTAINTY AND FRANCHISING
     POLICY 194
     Design Constraints 196
     An Example of Modularized System Design 198
     Relation Between System Design and
        Franchise Structure 206

IV. APPROACH TO FRANCHISE DECISIONS 210
     The Planning Process 210
     The Franchising Process 213
     Maintaining a Continuing Effort 214

Appendix
A. FCC DEFINITION OF CLASSES OF CABLE TELEVISION
    CHANNELS 217
B. FCC TECHNICAL STANDARDS 218
C. SAMPLE FRANCHISE PROVISIONS ON TECHNICAL
    STANDARDS AND PERFORMANCE MONITORING 221
D. SYSTEM RELIABILITY AND MAINTAINABILITY 224
E. FCC DEFINITION OF A CABLE TELEVISION SYSTEM 225

REFERENCES 226

Chapter 4. **Citizen Participation in Planning** 227
    Robert K. Yin

Section
  I. INTRODUCTION 229
     Participation: The Citizen's View 229
     Participation: The Municipal Official's View 230

  II. SETTING THE SCENE 233
     Early Developments: Reform to Renewal 233
     Midcentury: The Feds Take Over 236
     Lessons for Cable Television 239

  III. COMMUNITY ISSUES IN THE PLANNING OF CABLE
       TELEVISION SYSTEMS 242
     Ownership 242
     Cable System Geography 243
     Cable Subscription Fees 245
     Cable Services 246
     Monitoring Cable Operations 246

  IV. PLANNING FOR CABLE TELEVISION WITH CITIZEN
       PARTICIPATION 248
     When Participation Might Begin 248
     What Forms Citizen Participation Might Take 249
        Surveys 250
        Conferences 253
        Ad Hoc Committees 253
        Delegated Groups 254

  V. CONCLUSION AND RECOMMENDATIONS 255

Chapter 5. **Citizen Participation After the Franchise** 257
    Monroe E. Price and Michael Botein

Section
  I. INTRODUCTION 259

  II. BACKGROUND OF THE FCC RULES 260
     FCC Rules on Cable Use of Broadcast Television Signals 260
     FCC Rules on Use of Nonbroadcast Cable Channel Capacity 261
     FCC Rules on Federal, State, and Local Regulatory
       Relationships 262

  III. THE CERTIFICATE OF COMPLIANCE 264

  IV. GETTING ACCESS TO THE APPLICATION FOR A
       CERTIFICATE OF COMPLIANCE 266

  V. ANALYZING AN APPLICATION 267

VI. AN ANALYTIC APPROACH TO CERTIFICATE APPLICATIONS .. 269
    Are There Violations of Federal Standards? 269
    Supporting Franchise Provisions that Exceed
        Federal Standards 271
        Additional Community Channels 272
        Excessive Franchise Fees 274
    Should the FCC Add New Requirements? 275
    A Special Case: The Certificate Application and
        Broadcast Signals 276
    What Remedies Can the Community Group Seek? 277

VII. SPECIAL WAIVER AND RULEMAKING PROCEDURES 279
    Reaction of Local Governments and Community Groups
        to Petitions for Special Relief 279
    Petitions for Special Relief by Local Governments
        and Community Groups 281
    Local Government and Community Group Use of
        Rulemaking Proceedings 282
    Sources of Aid for Local Governments and Community
        Groups in Special Relief and Rulemaking Procedures 283

VIII. CONCLUSION 284

Index 285

Selected Rand Books 291

The Contributors 293

Preface

THIS book is the second of four volumes that present the results from a Rand Corporation study of cable television. The study was supported by a grant from the National Science Foundation to the Rand Communications Policy Program. A grant from the John and Mary R. Markle Foundation also aided completion of this volume.

Rand began its research on cable television issues in 1969, under grants from The Ford Foundation and The John and Mary R. Markle Foundation. The central interest at that time was federal regulatory policy, still in its formative stages. Rand published more than a dozen reports related to that subject over the next three years. This phase of Rand's concern ended in February 1972 when the Federal Communications Commission issued its *Cable Television Report and Order.*

The *Report and Order* marked the end of a virtual freeze on cable development in the major metropolitan areas that had persisted since 1966. It asserted the FCC's authority to regulate cable development, laid down a number of firm requirements and restrictions, and at the same time permitted considerable latitude to communities in drawing up the terms of their franchises. It expressly encouraged communities to innovate, while reserving the authority to approve or disapprove many of their proposed actions.

The major decisions to be made next, and therefore the major focus of new cable research, will be on the local level. These decisions will be crucially important because cable television is no longer a modest technique for improving rural television reception. It is on the brink of turning into a genuine urban communication system, with profound implications for our entire society. Most important, cable systems in the major markets are yet to be built, and many cities feel great pressure to begin issuing franchises. The decisions shortly to be made will reverberate through the 1980s.

Aware of the importance of these events, the National Science Foundation asked Rand in December 1971 to compile a cable handbook for local decisionmaking. The handbook, Volume I in this series, presents basic information about cable television and outlines the political, social, economic, legal, and technological issues a community will face. This book (Volume II) explores cable technology and issues of planning, franchising, and regulating a cable system in more detail. Other volumes discuss the federal regulations that apply to cable (Volume III) and the uses of a cable system for education, local government services, and public access to television (Volume IV).

The entire series is addressed to local government officials, educators, community group members, and other people concerned with the development of cable television in their communities. It also is intended as text and reference material for college and university classes in communications.

The study director, Walter S. Baer, served as editor for the series. Other contributors to this volume include Leland L. Johnson, Manager of Rand's Communications Policy Program; Robert K. Yin, Rand research psychologist; and consultants Michael Botein, then Assistant Professor of Law, University of Georgia; Carl Pilnick, President, Telecommunications Management Corporation, Los Angeles, California; and Monroe E. Price, Professor of Law, University of California, Los Angeles. Daniel Alesch, Edwin Deagle, Nathaniel Feldman, Norman Hanunian, Hans Heymann, Marvin Lavin, Howard Liberman, William Lucas, Jacob Mayer, Bridger Mitchell, Victor Nicholson, Steven Rivkin and Peter Szanton provided helpful comments and suggestions on various chapter drafts.

The views expressed in this book are those of the authors and do not necessarily reflect the opinions or policies of the National Science Foundation or the John and Mary R. Markle Foundation.

July 1973

Walter S. Baer
Santa Monica, California

xii

Chapter 1

A Guide to the Technology

Carl Pilnick and Walter S. Baer

I. INTRODUCTION

Cable television is a communication system that distributes television signals and other information by wire rather than through the air. More than twenty years old, it is just now turning from infancy to adolescence. The changes are only partly technological, but an evolving technology provides the base for cable's economic development and its future usefulness to society.

Cable began as a service to communities with inadequate off-the-air television reception, either because TV transmitters were too far away or because mountains, tall buildings, or other obstacles stood in the direct broadcast path. Cable provided a better television picture and usually more channels than viewers could receive directly off the air.

Today, most of these small towns and rural areas already have cable television. The urban centers and their surrounding suburbs remain as the principal areas of the United States yet to be wired; but because most of these areas already have good broadcast television reception, cable must offer a new range of services to gain subscriptions.[1] Cable has also been slow to penetrate the cities because of regulatory policies of the Federal Communications Commission (FCC). From 1966 to 1972, the FCC prohibited cable systems in the 100 largest television markets[2] from carrying television stations outside their market areas. This ban was partially lifted by new FCC regulations that became effective March 31, 1972. These rules also set up new requirements for major market cable systems, including a minimum capacity of twenty channels, allocation of three access channels, local program origination for systems with more than 3500 subscribers, and some capacity for two-way communications.[3]

Thus, both market factors and new regulations demand that new cable systems be constructed differently from those of the past. Building in additional communications capacity for new services is the principal design change. Before we turn to these developments, however, it will be helpful to review the basics of cable system design.

---

[1] R. E. Park, *Prospects for Cable in the 100 Largest Television Markets,* The Rand Corporation, R-875-MF, October 1971.

[2] Also called the "major markets," they are listed in Appendix A of Walter S. Baer, *Cable Television: A Handbook for Decisionmaking,* Crane, Russak & Co., New York, 1974.

[3] The 1972 FCC cable regulations are outlined in Baer, op. cit. They are reproduced in full and discussed at greater length in Steven R. Rivkin, *Cable Television: A Guide to Federal Regulations,* Crane, Russak & Co., New York, 1974.

CABLE TELEVISION DISTRIBUTION VERSUS BROADCASTING

A coaxial cable, pictured in Fig. 1, provides an electronic "information highway," no different in principle from a telephone wire or a wireless communication link such as broadcast radio or television. In each case information is sent as a varying electrical signal generally superimposed on a high frequency "carrier."[4]

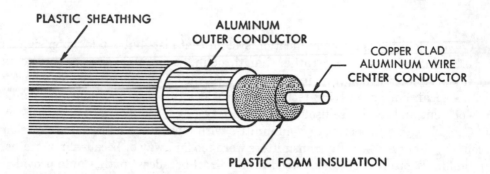

PLASTIC SHEATHING

ALUMINUM OUTER CONDUCTOR

COPPER CLAD ALUMINUM WIRE CENTER CONDUCTOR

PLASTIC FOAM INSULATION

Fig. 1—A typical coaxial cable

The higher the frequency of the composite signal (information plus carrier), the more information the system can transmit. All signals lose strength, however, as they travel from the transmitting point—an effect known as attenuation—and high-frequency signals are attenuated more than low-frequency ones. Consequently, there is a practical limit to the range of frequencies ("bandwidth") any electronic communication link can carry over a given distance. Broadcast television stations in the VHF band (channels 2-13) often provide good-quality signals fifty miles from their transmitters if no obstacles stand in the way. Higher-frequency, UHF stations (channels 14-83) usually cover smaller areas.[5] Cable systems are much more limited in range, as discussed in Sec. II.

Cable systems can deliver more television channels, however, since signals on the cable are less subject to adjacent channel interference than are those transmitted through the air. Broadcast television stations must be separated in frequency in

---

[4] Information can be carried either as an "analog" signal continuously varying in amplitude (AM) or frequency (FM), or as a "digital" signal composed of a series of discrete pulses. Television sound and picture information now is sent in analog form; digital signals are more appropriate for computer data and messages. The use of digital transmission over the telephone network is increasing rapidly for voice conversations as well as for data traffic.

[5] The FCC defines three classes of broadcast signals:

Principal City Service—Satisfactory picture quality expected at least 90 percent of the time for at least 90 percent of the receiving locations.

Grade A Service—Satisfactory picture quality expected at least 90 percent of the time for at least 70 percent of the receiving locations.

Grade B Service—Satisfactory picture quality expected at least 90 percent of the time for at least 50 percent of the receiving locations.

Maps showing Grade A and Grade B contours for VHF and UHF television stations in the United States can be found in the *Television Factbook*, Stations Volume, published annually by Television Digest, Inc., 1836 Jefferson Place, N.W., Washington, D.C. 20036.

a given area; for example, a community cannot receive both channels 12 and 13 over the air. In contrast, a well designed cable system can deliver adjacent television channels without appreciable interference. The abundance of channels, along with cable's ability to serve areas where over-the-air reception is inadequate, is the present basis of cable's technical advantage over broadcast television. In addition, cable systems can select audiences for pay-TV or other programs, and can provide two-way response communications from the viewer to the program source. These features are discussed in later sections, of the report.

HOW A BASIC CABLE SYSTEM WORKS

Figure 2 illustrates a conventional cable TV system designed to distribute broadcast television programming. Its components include a tower and antennas to receive broadcast television signals, a "headend" to process them and add other signals, and the cable distribution network that carries the signals to subscribers' TV receivers. Distribution is on a "party-line" basis from a single origination point (the headend), with each subscriber having access to exactly the same programming.

Towers and Antennas

Antennas to receive broadcast TV signals usually are located on one or more high towers. This is because broadcast signals are blocked by the curvature of the earth and are only partially reflected by the atmosphere. Thus, a sufficiently strong signal will be received only where there is an unrestricted line-of-sight path between a TV station's transmitter and the cable system's antenna.

Cable systems in rural areas place towers on a mountain top or other high ground. For urban cable systems, the roof of a tall building may be most suitable. The taller the building, the less need for a high tower. Tower heights may vary from 20 to 30 feet for some systems to 100 to 500 feet for others.

A separate antenna is used for each TV station received, so that it may be tuned to the station broadcast frequency and mechanically aligned to pick up the strongest signal. Where off-the-air signals are especially weak, a preamplifier may be used for each channel. This unit is mounted as close to the antenna itself as possible, usually on a mast of the tower, so that it can boost the desired signal before additional noise is introduced.

In areas with few local TV stations, the cable system operator may wish to bring in distant signals to provide his subscribers with a greater diversity of programs. These distant signals would originate from stations too far away to be received directly by an antenna at the cable system's tower. Consequently, they must be relayed to the cable system's headend, either by microwave transmission (at frequencies specially licensed by the FCC for this type of relay) or by a large-diameter coaxial cable. The choice depends upon distance, the number of signals to be carried,[6] availability of microwave frequencies, and cost.

---

[6] The FCC permits major market cable systems to import 2 or 3 commercial signals. Other cable systems can import more. See Baer, op. cit., or Rivkin, op. cit., for details.

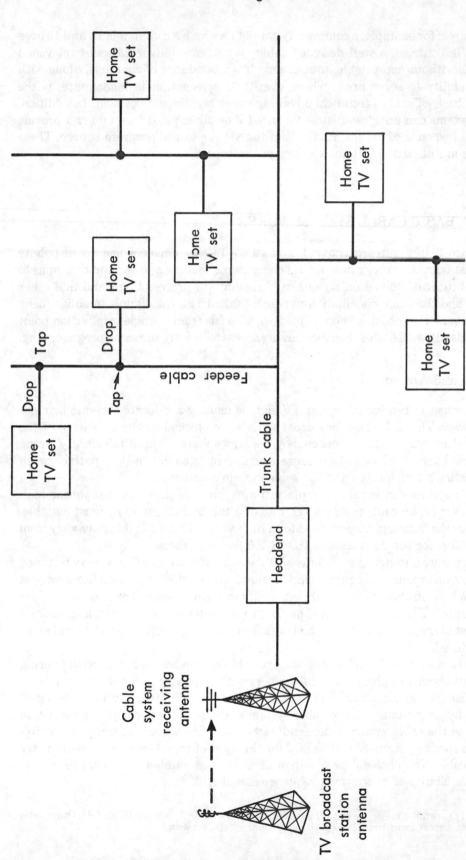

Fig. 2—A basic cable television system

Figure 3 illustrates distant TV signal relay by a single microwave link, which can cover a distance of 20 to 30 miles. Additional links would be used for longer distances, with each receiving the signal and retransmitting it to the next.

Headend

From the antenna preamplifier, if used, each broadcast signal is connected by cable to the headend facility, usually located in a small building near the tower site. For urban systems, this may be an office in the building on which the receiving tower is mounted. The headend contains the electronic equipment necessary to process signals for distribution on the cable network. A variety of functions are required:

- Undesired signals outside the frequency band of each channel must be filtered out.
- Some channels must be "translated" in frequency before being sent through the cable. UHF stations, for example, are converted to an otherwise unused VHF channel so that they can be selected for viewing at the VHF tuner of each subscriber's TV set. VHF stations also may have to be translated because of local interference problems.
- The TV information from distant signals imported by microwave link must be separated from the microwave transmission carrier. This is known as "demodulation."
- Demodulated signals and all video signals that originate within the cable system must be combined with a carrier frequency to match a channel in the standard VHF TV band. This is known as "modulation."
- All signals must be "mixed" or combined into a composite signal, and then amplified before being distributed on the cable. Mixing also includes adjustment for the fact that the higher frequencies will undergo more attenuation losses in the cable and would, unless compensated, emerge as weaker signals. "Equalization," which in effect amplifies high-frequency signals more than low-frequencies, adjusts for the differential losses.

Equipment to perform these functions is available from a variety of manufacturers, with performance characteristics and prices generally very competitive. New cable systems now being planned or franchised, however, will require other kinds of headend equipment now available only in developmental form or, in some cases, not available at all. This includes equipment for private channel viewing, headend interconnection, and two-way communications. Consequently, the complexity and cost of headend facilities for large cable systems are bound to increase in the future.

Facilities for Local Origination

In the 1950s, many cable operators began to originate their own programming. The simplest and least costly forms of local origination are the so-called "automated services." For example, a TV camera might be permanently focused on a clock and weather indicator and left unattended. The picture would be sent out over an unused channel of the cable system to provide subscribers with continuous time and weather information. Stock market or news information can be displayed in similar fashion. Equipment now is available that can take the electrical output of a stock or news ticker and convert it line-by-line for video display on conventional TV sets.

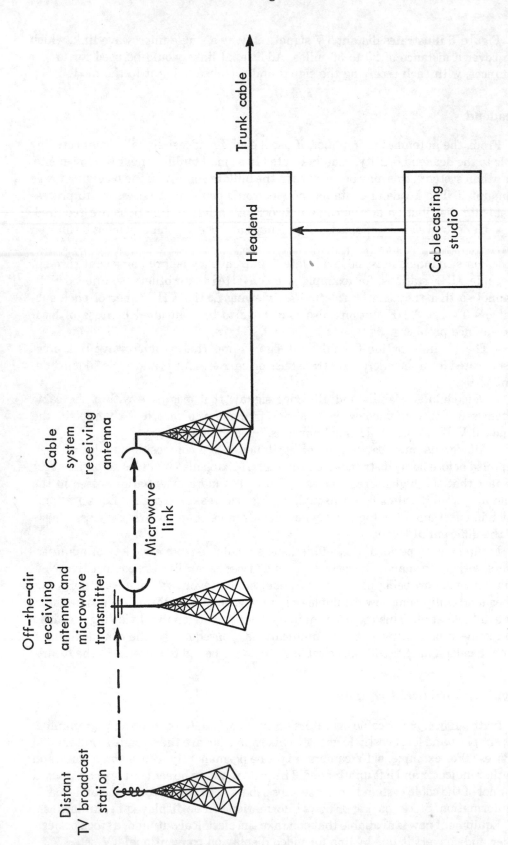

Fig. 3—Cable system with distant signal transportation and local origination

According to *Television Factbook,* about one-third of all U.S. cable systems provide some automated services.[7] Many cable operators in small communities contend this is all the local programming they can afford. Other operators, however, have built studio facilities to distribute live, filmed, or taped programs to their subscribers. This requires much of the same equipment—TV cameras, lights, videotape recorders, film chains, and so forth—that TV broadcast stations use. In effect, such local program origination gives cable subscribers another TV station to choose from. The cable industry calls it "cablecasting," as opposed to broadcasting. The main difference is that a cable operator can identify his audience more precisely and, theoretically at least, adapt the cablecast programs to particular local needs or desires. As of March 1972, about 20 percent of U.S. operating cable systems had the capability to provide full cablecasting service over at least one channel.

Cablecasting studios typically use smaller and less expensive TV cameras and videotape recorders than do their broadcast counterparts. They also are more restricted in studio space, set design, lighting facilities for studio audiences, and program production personnel. Today, cable studio facilities range widely in cost from perhaps $10,000 to $25,000 at the low end, to $250,000 to $500,000 for the largest systems. Cablecasting facilities and equipment are discussed further in Sec. III.

Many cablecasting facilities are located in the same building as the headend. When they are, it is easier to connect studio program signals into the cable distribution network. In other systems, origination facilities may be some distance from the headend. A studio should be convenient to those involved in program production, which generally implies a location in the more populated downtown area of a community, while the headend must be close to the receiving tower, whose location is determined by both signal strength and real estate constraints. If the two are at different locations, a separate cable carries cablecast programs from the studio to the headend.

Cable Trunk

The main cables that carry signals from the headend are called *trunk* cables. They are usually 1/2 or 3/4 of an inch in diameter, but may be as large as 1 or even 1-1/4-inches. The larger diameter cables are used for longer distances, since they attenuate signals less. They are more expensive, of course, and also difficult to install because they are heavy and less flexible.

Coaxial cables are either strung on utility poles or placed in underground ducts. In an aerial system, the cable operator rents space on existing utility poles, whose number and location usually are adequate for the cable system. However, the telephone or power utility must rearrange its own wires and otherwise make the poles ready for cable installation. Although this process naturally takes time and costs money, cable operators often complain bitterly about undue delays and excessive charges for obtaining their "pole rights," particularly from telephone companies who might be present or future competitors. As one example, the Kern Cable Company of Bakersfield, California described its problems in a letter to the local County Board of Supervisors:

[7] *Television Factbook,* Services Volume, 1972-73 edition, p. 75a.

The primary factor in Kern Cable's inability to meet its construction costs budget can be attributed to an unexpected change in Pacific Telephone Company's policies relative to the attachment of CATV cables to its poles. . . . Whereas rearrangement costs were originally budgeted at $90,000, the final Telephone Company bills totalled $270,000, an unanticipated expenditure of $180,000.[8]

More recently, the General Telephone Company of California has proposed to double its pole rental rates to cable systems.[9] Since obtaining utility "pole rights" and pole rearrangements often becomes a bottleneck in cable system construction, the FCC is currently considering setting federal regulations in this area. The local franchising authority also may want to monitor the negotiations between the cable operator and the utility that owns the poles. Both parties, of course, are obligated to bargain in good faith. But it may also help if the city clearly expresses its interest in a reasonable and expeditious settlement after a cable franchise is awarded.

Many communities today prefer underground utility construction. Where existing utilities already are underground, the cable system operator normally follows the same practice. The cost is two to twenty times that of aerial construction, depending on whether existing utility ducts have enough room to accommodate the new cables, problems of digging in the city streets, the necessity for protecting against water seepage and corrosion, and so forth.[10] As with pole rights, negotiations with utilities for duct space may be difficult and time-consuming.

Since each city has its own network of subsurface ducts, techniques for underground cable installation are less standardized. Amplifiers are often placed in aboveground housings for easier maintenance and repair. The cable itself is accessed through normal utility manholes, usually with difficulty. As a result, routine maintenance and replacement costs, as well as initial construction costs, are much higher in underground systems.

The actual cable construction and installation must conform to a variety of codes and regulations, such as the National Electric Safety Code, Underwriters' Laboratories standards, state and city building codes, and the practices required by the utility companies from which pole or duct space is rented.

Amplifiers

Television signals are attenuated throughout the cable system. A signal at the upper frequency limit of current cable TV systems will lose about half its power in a 200-foot length of 1/2-inch trunk cable. Consequently, amplifiers must be placed throughout the cable route to build the signal strength back up to usable levels.

Trunk amplifiers are particularly critical components. Each one degrades the TV signal quality slightly, and their effect in series is cumulative. Thus, there is a practical limit to the number of amplifiers that can be cascaded before the signal quality becomes unacceptable. Where cable competes with good off-the-air reception,

---

[8] Letter from Kern Cable Company to Kern County Board of Supervisors, February 18, 1970.

[9] *CATV,* April 30, 1973, p. 3.

[10] Gary Weinberg, *Cost Analysis of CATV Components: Final Report,* Resource Management Corporation, Bethesda, Maryland, 1972, provides useful guidelines to the costs of underground cable construction.

the limit may be 20 to 30 amplifiers in cascade, corresponding to a cable run of 5 to 10 miles from the headend.[11]

Trunk cable distances must be carefully related to the geography of the area served. To minimize trunk runs, many current systems use a "hub" concept, with the headend at the center of a group of trunk cables strung radially like spokes on a wheel. A single hub theoretically can serve a circle of up to 100 square miles, but municipal boundaries and geographic constraints often limit the effective service area to 50 square miles or less. Larger systems require multiple headends.

Feeder Cable

When a trunk cable passes a residential street or other area of concentrated subscriber density, a smaller distribution or *feeder* cable is used to distribute signals from the trunk to that area. Feeder cables are similar in construction to the trunk, but of smaller diameter (0.412-inch is a popular size).

The trunk and the feeder are connected through a "bridging" amplifier, which electrically isolates the trunk to prevent electrical interferences from degrading the trunk signals. "Line extender" amplifiers boost signals within feeder lines so that more subscribers can be served. Since feeder cable is cheaper than trunk, cable systems generally try to maximize the feeder-to-trunk ratio for lowest costs.

Figure 4 depicts an aerial feeder cable installation. The coaxial communications cable is lashed to a steel "strand" or "messenger" cable for mechanical support. A line extender amplifier appears above the lineman's head. Next to it is a "tap" that connects subscriber "drop" lines to the feeder cable.

Subscriber Taps and Drops

A small *drop* cable brings the signals from the closest feeder line to the home TV set. The drop cable is generally 1/4 to 1/3 of an inch in diameter. A coupler, or *tap*, connects the drop to the feeder cable. Conventional taps offer low resistance to signals flowing from the feeder into the home, but high resistance to reverse signals. This reduces the possibility of interference emanating from subscribers' homes and entering the cable network. Special two-way taps must be used for return communications from the home.

At the subscriber's home, the drop cable may connect to a small transformer that matches the characteristics of the cable to the input of the TV set. Many new cable systems use set-top converters to provide more than 12-channel capacity (see Sec. II). If a converter is employed, the drop cable will connect to its input. The subscriber also may want a switch to connect his set back to a rooftop antenna should the cable system fail.

---

[11] Standard trunk amplifiers are designed to be placed about every 1500 feet for a 1/2-inch cable or every 2000 feet for a 3/4-inch cable. A 25-amplifier chain thus corresponds to a cable trunk run of approximately 7 and 9.3 miles, respectively. However, since cable trunk lines generally follow rectangular street grids, the effective radius serviced from a single headend is only about 5 to 7 miles.

Courtesy: National Cable Television Association

Fig. 4—Typical aerial, single-cable installation

EXTENDING THE RANGE OF CABLE SERVICES

Building the conventional, one-way cable system described above typically costs about $60 to $75 per home passed if most of the construction is aboveground. Assuming 50 percent of households subscribe for service, the system's initial construction cost is $120 to $150 per subscriber. Many CATV systems that distribute only broadcast TV signals have been built for considerably less.

New cable systems in the major markets, however, are expected to provide more than broadcast TV redistribution. The new FCC rules require them to provide 20 or more channels, produce local programming "to a significant extent" if they have more than 3500 subscribers,[12] provide channels for public access, education, and local government services, make additional channels available on a leased basis for pay TV and other uses, and provide the capability for eventual two-way services. As shown in Table 1, most U.S. cable systems offer none of these added services today. Providing them requires new technical approaches and undoubtedly will raise the cost of cable construction above the figure of $150 per subscriber used in the past. The technology for added cable capacities and services is the subject of the remaining sections of this report.

[12] About 17 percent of U.S. cable systems had more than 3500 subscribers as of March 1972, according to *Television Factbook*.

Table 1

PRESENT STATUS OF CABLE TELEVISION SERVICES

Service Category	Type of Communication	Present Availability
Distribution of broad-cast television programs	One-way, headend to all subscribers	Operational on all cable television systems
Local cablecasting	One-way, headend to all subscribers	Operational on about 20% of U.S. systems; required of all systems with more than 3500 subscribers
Public access, educational, and government channels	One-way, headend to all subscribers	Operational on a few systems; required of all new major market systems
Pay TV or private channel programming	One-way, headend to certain subscribers (two-way useful, but not required)	In prototype form on a few systems
Subscriber response services	Two-way, data response from subscribers to headend	Under field test on a few systems but not yet operational; "technical capacity for nonvoice return communications" required of all new major market systems
Information retrieval, document delivery, and other "new services"	Two-way data, voice, and video between sub-scribers and headend, and possibly among subscribers	Under development

II. COMMUNICATION CAPACITY OF CABLE SYSTEMS

The communication capacity of a cable system has been measured by the number of television channels it could deliver simultaneously to subscribers. The earliest cable systems carried 3 or 5. Most systems built in the last decade have had a 12-channel capacity. Since March 31, 1972, FCC regulations require at least 20 channels for new construction in the major markets.

Channel capacity can be somewhat misleading as a measure of the nonbroadcast services a cable system can deliver. More generally, a system's communication capacity is measured by its bandwidth in cycles per second (or in the more modern units of "hertz," abbreviated Hz). Each U.S. standard television channel requires a large frequency bandwidth of 6,000,000 hertz, usually stated as 6 Megahertz, abbreviated 6 MHz. Thus, the FCC's 20-channel requirement actually means a usable bandwidth of 20 x 6, or 120 MHz.

In contrast, telephone channels need only 3000 to 4000 hertz, or 3 to 4 Kilohertz (KHz). Data channels for sending short messages or alarm signals may require only about 100 Hz. A 20-channel cable system thus in principle could carry 30,000 to 40,000 telephone messages instead, or more than one million fire and burglar alarm channels. Certain practical inefficiencies would reduce these numbers, but data messages in 10,000 or more homes could be sent and received in a portion of a single 6 MHz channel.

Technical limitations on channel capacity are set principally by cable amplifiers. Current amplifiers are limited to a usable bandwidth of about 300 MHz. While this theoretically is equivalent to 50 television channels, interferences among channels give a practical limit of about 25 to 35 channels for each cable.[1]

THE BASIC 12-CHANNEL SYSTEM

Most cable systems previously were designed for 12 channels, however, to match the 12 channels of the standard VHF tuner on television receivers (channels 2-13). If a cable carried more than 12 channels, a subscriber could view them only

---

[1] Engineers find it hard to agree on a precise channel limit, since it depends on the signal quality one is willing to accept, the particular amplifier and cable system design, and environmental factors such as the temperature changes the amplifier is subject to. The 25-to-35 channel range represents some upper limit to today's state of the art.

with a special converter attached to the TV set. Using the UHF tuning capacity of the receiver would not work, because UHF frequencies starting at 470 MHz are too high for the cable system to carry directly. Television channel frequency assignments are shown in Table 2.

As a result, any UHF station picked up by a cable system's antennas must be translated in frequency (down-converted) at the headend before being inserted onto the cable. An otherwise unused VHF channel is the best choice, since subscribers can then tune it in directly on their TV sets. Thus, Channel 28 might be translated to the frequency band for Channel 12, if Channel 12 is not broadcasting in that region, and be viewed by tuning directly to Channel 12.

Some translations are dictated, therefore, by the need for UHF-to-VHF conversion. Still other translations are necessary because of the phenomenon known as "direct" or "on-channel" interference. In most metropolitan areas, the local TV broadcast stations provide a strong signal at the antenna connections of TV sets within 5 to 10 miles even without an external antenna. The receiver picks up this broadcast signal in addition to the signal delivered by the cable on the same channel. Because television signals travel slightly slower through cable than they do through the air, the cable signal arrives a small fraction of a second later than the off-the-air signal. The tiny difference is enough to cause a "ghost" image on the TV receiver that can make the picture unacceptable (Fig. 5).

Consequently, if a community has three strong VHF broadcast stations, say channels 2, 4, and 7, those three channels probably cannot be transmitted on the cable at their usual frequencies. Instead, they would have to be translated in fre-

Table 2

FCC-ASSIGNED TELEVISION CHANNEL FREQUENCIES

Channel Number	Assigned Frequency (Mhz)	Comment
2	54–60	Lowest VHF channel
3	60–66	
4	66–72	
---	72–76	Not assigned for broadcast
5	76–82	
6	82–88	
---	88–108	FM band
---	108–174	Not assigned for broadcast
7	174–180	
8	180–186	
9	186–192	
10	192–198	
11	198–204	
12	204–210	
13	210–216	Highest VHF channel
---	216–470	Not assigned for broadcast
14	470–476	Lowest UHF channel
15–82	476–884	6 Mhz per UHF channel
83	884–890	Highest UHF channel

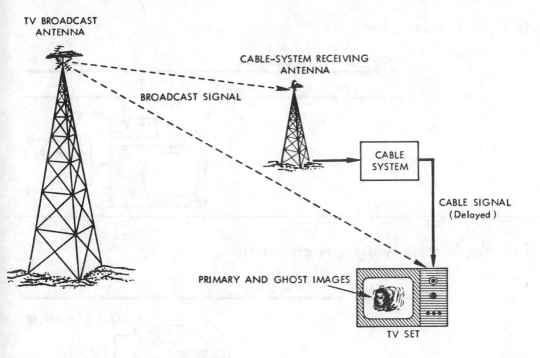

TV BROADCAST
ANTENNA

CABLE-SYSTEM RECEIVING
ANTENNA

BROADCAST SIGNAL

CABLE
SYSTEM

CABLE SIGNAL
(Delayed)

PRIMARY AND GHOST IMAGES

TV SET

Fig. 5—Direct signal pick-up interference

quency, and perhaps delivered as Channels 3, 5, and 8. Cable channels 2, 4, and 7 might still be usable for automated services and other nonbroadcast applications, however, where the interference is less damaging to the displayed information.

Large cities such as New York and Los Angeles may have as many as seven local VHF stations. As a result, seven cable channels may be unusable for television rebroadcast, and a 12-channel system, without converters, could not even deliver all the stations the subscriber could receive directly off the air. For major markets, the 12-channel cable system represents only a historical precedent. It is technologically inadequate even without the FCC 20-channel requirement.

EXPANDED CHANNEL SYSTEMS

Expanding channel capacity requires a new cable system design. As illustrated in Fig. 6, the three principal choices are multiple cables, converters, and switched systems.

Multiple Cable Systems

If one cable can deliver 12 directly selectable channels to the TV set, an obvious solution to expanding capacity is to use two or more cables. Figure 6(a) shows two trunk cables, each of which can carry up to 12 signals. The subscriber is furnished

18

(a) Dual cable system

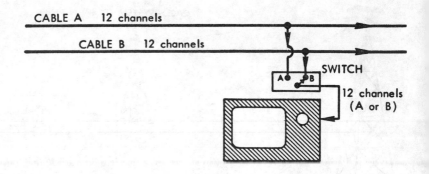

(b) Single cable system with converters

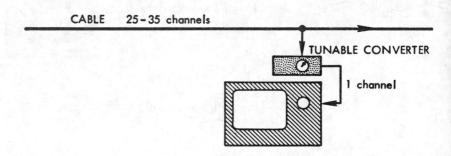

(c) Switched system

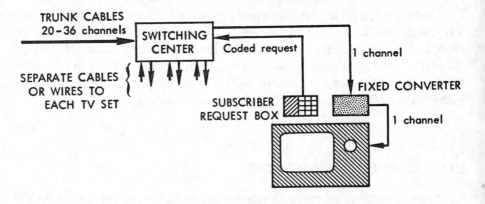

Fig. 6—Three methods of expanding channel capacity

a two-position "A-B" switch (carefully designed for switching at VHF frequencies) and can select which of the two cables is connected to the TV set at any particular time.

This doubles the viewing capacity. It does not address the problem of on-channel interference. If Channel 4 is subject to such interference, it cannot be used for either cable. Consequently, instead of the nominal 24-channel capacity, a community with four strong over-the-air broadcast stations would have only 16 usable channels on a dual cable system. This could still satisfy the FCC's 20-channel requirement, since all 20 channels need not be used for broadcast television, but it may not be what the community wants. One solution is to connect the subscriber's switch directly to the tuner inside his set with shielded cable, but this requires modification of each TV set subject to on-channel interference.

Dual (or multiple) cable systems are obviously more expensive than single cable systems—not twice as much, since the added installation costs are small, but about 50 percent more. This may or may not be competitive with other ways to expand channel capacity and must be examined on a case-by-case basis. The principal advantage of the multiple cable approach is its simplicity. It eliminates converters, which are problem components, and thus makes the system more reliable. The cable can carry signals at standard VHF frequencies selected to minimize interference among channels. Finally, if each cable is designed initially for expanded channel capacity and return communications, a dual cable system doubles the overall capability for two-way or other new services.

Systems with Converters

A converter changes a nonstandard frequency channel to a VHF channel that can be tuned directly on the subscriber's TV set. Some older style converters, called block converters, translated an entire block of 12 channels from a higher frequency range to the standard VHF band. Most present converter systems use tunable converters, shown in Fig. 6(b). In effect, the converter replaces the standard TV set tuner and provides more channel positions. Channels may be selected with a dial like that of standard tuners, a slide lever, or push buttons; Fig. 7 illustrates three current models.

Carrying more than 12 channels on a single cable requires higher quality amplifiers and more careful system design, both of which cost more money. The extra channels in a converter system are carried on the cable at frequencies between channels 6 and 7 (known as the midband) and above channel 13 (known as the superband). The industry today designates nine midband and thirteen superband channels below 300 MHz, as shown in Table 3. If all were usable in addition to the twelve standard VHF channels, a single cable could carry 34 6-MHz channels. Today's converters, however, are designed for a maximum of 25 to 30 video channels in the VHF, mid-, and superbands. Seven other channels below channel 2 (the sub-band) are usually reserved for two-way or other new applications.

The converter changes the frequency of a selected channel to a standard VHF channel frequency that is unused for broadcasting in the community. The TV set tuner is set permanently to that channel, and all selection is performed at the converter. Unlike conventional TV tuners, its input is shielded from off-the-air

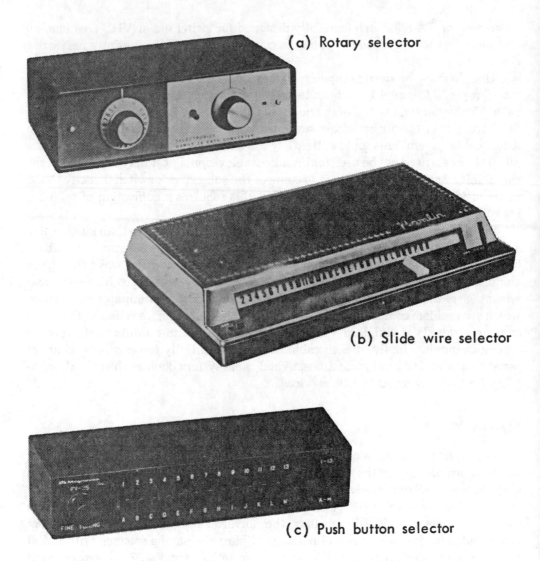

(a) Rotary selector

(b) Slide wire selector

(c) Push button selector

Fig. 7—Three types of cable television converters

Table 3

NONSTANDARD CABLE CHANNELS

Channel Nomenclature	Frequency (Mhz)
Sub-band	
T7	5.75–11.75
T8	11.75–17.75
T9	17.75–23.75
T10	23.75–29.75
T11	29.75–35.75
T12	35.75–41.75
T13	41.75–47.75
Midband	
A	120–126
B	126–132
C	132–138
D	138–144
E	144–150
F	150–156
G	156–162
H	162–168
I	168–174
Superband	
J	216–222
K	222–228
L	228–234
M	234–240
N	240–246
O	246–252
P	252–258
Q	258–264
R	264–270
S	271–277
T	277–283
U	283–289
V	289–295

signals. The converter thus completely eliminates on-channel interference, since its output will never be at the same frequency as a strong local station.

However, converters introduce other interference and picture degradation difficulties. Many converters respond inadequately to variations in signal strength and are overloaded by strong input signals. This causes picture distortion. The frequency of the converter oscillator can drift with temperature and time, resulting in inaccurate frequency conversion. Channel selectivity—the ability to distinguish sharply between adjacent channels—is sometimes poor. And because more frequencies are carried on the cable, more interference problems among channels arise.

These problems are due more to an emphasis on low cost in converter design than to intrinsic technical limitations. Converters range in price from $35 to $40 in small quantities to $25 to $30 in lots of 1000 or more. Since a converter is needed for each TV set, a $30 unit cost may represent 15 to 20 percent of total system capital investment. Consequently, the pressures for low-cost converter design are great. Initial cost savings, however, may be outweighed over a few years by added service calls and subscriber complaints. Still, the set-top converter is today the most popular approach to providing a minimum 20-channel capacity. Even dual cable systems may need a converter to utilize channels subject to direct interference.

Switched Systems

Switched systems provide a completely different approach to expanded channel capacity by placing channel selection outside the subscriber's home, as in Fig. 6(c). The two principal switched systems under development are the Ameco DISCADE and the Rediffusion systems.[2] Both bring signals from a headend to switching centers that serve from twenty to several hundred subscribers. Two separate wires or cables run from the switching center to each subscriber receiver. One wire carries subscriber requests to the switching center, and the other returns the selected television signal.

Since each subscriber has his own link to the center and receives only one program at a time, a single, low-frequency band (perhaps 4 to 10 MHz) suffices to carry all programs. If a subscriber requests Channel 2 (normally 54 to 60 MHz), the signal would be down-converted at the switching center to 4 to 10 MHz before being sent to his receiver. Other channels would be similarly down-converted. The major advantage of this technique is that attenuation is less at these low frequencies, so that a smaller and cheaper coaxial cable, or even a twisted-wire pair, can be used for signal distribution to the home.

At the subscriber's home, a fixed-channel, tunerless converter translates the 4 to 10 MHz signal to a standard unused VHF channel. Since the same one-channel conversion is made every time, the converter can be a relatively simple device.

Switched systems are simple in concept and have advantages for certain applications. Their big disadvantage is the quantity and complexity of cabling they require. The switching centers service a relatively small number of subscribers, necessitating many centers for a large community. Moreover, each TV set requires a separate set of cables or wires to the switching center. In crowded urban areas, the cost of switching centers may be high and the cost of laying the necessary wiring underground may be prohibitive. One recent study estimated the Rediffusion system cost to be 50 percent more than that for a conventional, dual cable system.[3]

For smaller communities, where not too many switching centers are required and construction is mostly aerial, switched systems may well be attractive. They have not as yet, however, become a significant factor in current cable operations in the United States.

---

[2] Descriptions of these two systems can be found in J. E. Hickman and G. C. Kleykamp, "Multi-Cable Solution to Communications Systems Problems," Ameco, Inc., P. O. Box 13741, Phoenix, Arizona, 1971; and R. P. Gabriel, "Dial a Program: An HF Remote Selection Cable Television System," *Proceedings of the IEEE*, Vol. 58, No. 7, July 1970.

[3] John E. Ward, "Present and Probable CATV/Broadband-Communication Technology," Appendix A to the report of the Sloan Commission on Cable Communications, *On the Cable: The Television of Abundance*, McGraw-Hill, New York, 1971.

THE TV RECEIVER PROBLEM

None of the above techniques is an obvious "best solution" to the expanded channel problem. Systems with multiple cables eliminate converters, but do not prevent direct interference. Converters solve the direct interference problem completely, but introduce new possibilities for interference and picture degradation. Switched systems obviate both the direct interference and the converter problems, but seem too cumbersome and expensive for major market operations.

A principal design problem is that all cable components must be compatible with conventional TV receivers, which are today the weakest links in most systems. TV sets are designed to use signals received off the air, not from cable. They generally have poor selectivity, since adjacent broadcast channels are never assigned in any area, and therefore receivers need not have sharp tuning on one channel only. For all practical purposes, the 12-channel VHF tuner is really a 7-channel device under broadcast conditions.[4] The UHF tuner has been added (generally as an afterthought to meet the FCC requirements for UHF reception capability) as a separate device, with a different tuning method and performance characteristics. Shielding is inadequate to prevent strong broadcast signals from being picked up at the input stages of the set, even without an external antenna, and sometimes inadequate to prevent signals from being radiated out from the receiver.

Logically, a cable-compatible receiver could be designed that would ease the above problems and lower receiver costs as well. It would have a multichannel tuner built in (eliminating the need for a separate converter) and better input shielding (eliminating on-channel interference). Although TV set manufacturers are developing cable-compatible models, mass production must await agreement on standards and the emergence of a larger market of cable subscribers—perhaps 10 to 15 million households.

When special receivers are available, cable operators may want to lease them to subscribers as part of their regular cable service. Many existing cable franchises prohibit the operator from leasing or servicing TV sets, usually as a result of pressure from local TV retailers and repairmen. They argue that such leasing would give the cable operator an unfair competitive advantage. The operators contend that TV set leasing not only is fair, but also allows them to guarantee better quality reception and expedites the development of new services. Communities might well strike a middle course by neither forbidding TV set leasing in the cable franchise, nor permitting the operator to require receiver leasing as part of his overall service. The franchise also might permit the operator to make TV set modifications (as, for example, by connecting the cable directly to the tuner inside the set) where necessary to improve reception.

COMPARING ALTERNATIVES

Until cable-compatible receivers are available, a community's choice generally will be among a dual cable system, a converter system, or a hybrid of both. Table

[4] Some consecutively numbered VHF channels are not really adjacent in frequency, as shown in Table 2. Channels 4 and 5, or 6 and 7, can be assigned in the same area.

4 compares cable distribution costs per mile for several one-way system designs based on recent quotations from equipment suppliers (costs for systems with two-way transmission capability are shown in Table 10). Costs are also shown on a per-subscriber basis, assuming 40-percent penetration in each case.

Table 4

COST COMPARISON OF ONE-WAY CABLE SYSTEM DESIGN OPTIONS

System Characteristics	Cost per Mile, Aerial Cable Distribution Plant	System Cost per Subscriber, 40% Penetration[a]
Single cable, 12 channels	$4500-5000	$120-135
Single cable with converters, 20-30 channels	$5500-6500	$190-210
Single cable with converters, 20-30 channels plus shadow trunk	$6500-8000	$210-250
Dual cable, 12 channels each[b]	$6800-7500	$190-205
Dual cable, 20-30 channels each, with converters	$7800-9800	$250-300

[a]Assumes 100 homes per mile, converters at $35, dual cable switches at $4, headend costs of $10 per subscriber for 12-channel system, $15 per subscriber for other systems.

[b]Minus channels unusable due to on-channel interference.

One current hybrid approach is to install a single cable with converters, plus a second, "shadow" trunk—that is, an extra trunk without amplifiers or other electronic components. The shadow trunk is available when the demand for more channels or new services requires it. The shadow-trunk adds about 20 percent to distribution plant costs.

The other general approach is to build a full dual cable system—dual trunk, feeder, and drop lines—with converters where necessary. A dual cable system with converters may cost 30 percent more than a single-cable system and perhaps require a dollar a month greater subscription fee, but it provides more inherent capacity for future expansion. Whether this added flexibility is worth the extra cost is a question for each community to decide.[5]

---

[5] Unfortunately, the immediate and future benefits from additional capacity are much harder to quantify than the costs of providing it. Communities must therefore consider the specific ways in which added capacity will be used. For further discussion, see Chapter 3 of this book and other reports in this series.

III. CABLECASTING

Most cablecasting today takes place in the cable operator's studio, which, as mentioned in Sec. I, typically is a compact, less elaborate version of a TV broadcast studio. However, the development of low-cost, portable video cameras and tape recorders makes program production possible anywhere in the community. The easy availability and use of this equipment is responsible for much of the current interest in community program origination, or public access.[1]

STUDIO FACILITIES

Figure 8 shows, in block form, the major electronic equipment in a cablecasting studio that can produce live programs or distribute those already on film or videotape.

Live programming begins with at least one video camera that converts visual images into *video* electronic signals in a standard broadcast TV format. Two or more cameras give a more interesting balance among close-up, distance, and angle shots. At the same time, microphones convert the sound information into *audio* electronic signals. The video and audio are then connected into an array of "signal conditioning" equipment that performs three principal functions:

1. Mixing the audio and video in the right proportion.
2. Amplifying the signals to a level high enough for distribution.
3. Locating the audio and video carrier frequencies with respect to each other in conformance with broadcast TV standards.

A control console, usually located in an adjacent room, permits an operator to select which camera or microphone is on at any particular instant, and to vary audio and video levels for the best combination.

The combined signal, known as the "composite video," is then sent to a VHF modulator-amplifier. This component translates the video frequency band (0 to 6 MHz) up to a VHF channel that can be tuned in by a TV set. (For example, the

---

[1] See Baer, op. cit., Chap. 8, and Richard C. Kletter, "Making Public Access Effective," Chapter 1 of Polly Carpenter et al., *Cable Television: Developing Community Services*, Crane, Russak & Co., New York, 1974.

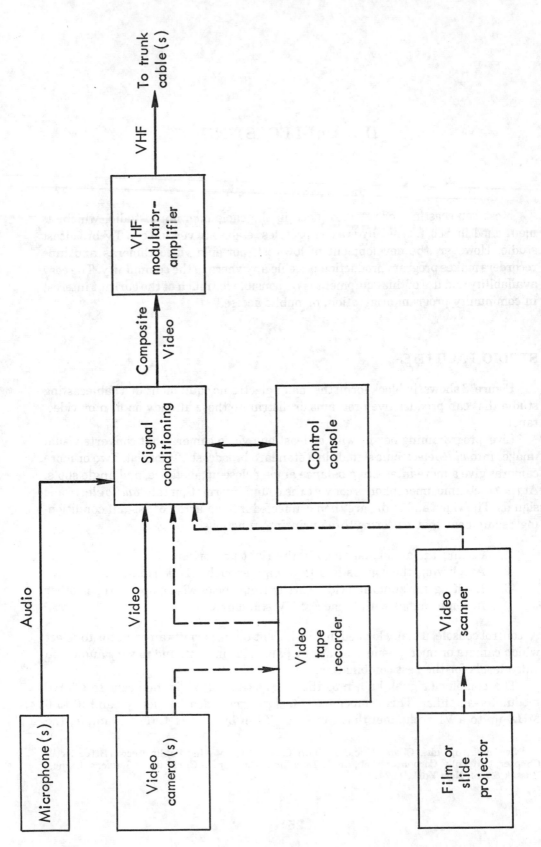

Fig. 8—Cablecasting studio block diagram

cablecast program could be carried on Channel 3's frequency band, 60 to 66 MHz, if this channel is not used for broadcast signals. A cable subscriber would then view the program by turning his VHF tuner to Channel 3.)

The outgoing signal is then amplified and sent through the cable network to subscribers. As described in Sec. II, cablecasting signals are combined at the headend with broadcast signals received off-the-air and those imported from distant cities. All signals are mixed and transmitted simultaneously, with a separate 6 MHz frequency band for each channel.

Many cablecast programs are transmitted from previously recorded tape or film. With a video camera and tape recorder, the cablecasting studio can tape programs, edit them, and play them back to subscribers at a later time. Most studios have two or more videotape recorder/player units, partly for backup in case one fails, but also to be able to reproduce (dub) a taped program by playing back from one and recording on the second.

A "camera chain" is required for programs that are on film strip or slides. This is basically a film or slide projector coupled to a special video camera. The projector will project the photographic image onto a high-intensity internal screen which is then scanned by the video camera. The optical image is converted into a video electronic signal, and then sent to the signal conditioning equipment in exactly the same way as a signal from a live or taped program.

Studios also require a variety of support equipment such as lights, camera dollies, microphone booms, power supplies, and television monitors. The total equipment cost can vary from a few thousand dollars for the simplest black-and-white installation to several hundred thousand dollars for a fully equipped color studio. Table 5 lists the costs for equipment packages offered by two representative manufacturers. The Audiotronics package represents a "very small" studio, while the RCA equipment is in the "small-to-medium" cost range. The cost of a "large" studio can easily reach $250,000 to $500,000 even without mobile equipment.

VIDEO CAMERAS AND TAPE RECORDERS

Video cameras and recorders deserve special comment, since they will often be used by nonprofessionals outside the cable operator's studio. Basically, a video camera combines parts of a motion-picture camera with an electronic scanning device. As in a film camera, a lens focuses an image and projects it on a plane inside the camera housing. This plane is actually the screen of an electronic "pickup" tube, which scans the image a point at a time and converts the shadings of light into a varying electrical signal. The signal is then converted into the form required to meet broadcast TV standards.

Two kinds of pickup tubes, the "Image Orthicon" and the "Vidicon," have been used in most cameras to date (the "Plumbicon," developed by North American Philips, has recently begun to offer competition to the Vidicon). The Image Orthicon is more sensitive and can operate at low light levels, but is relatively expensive. The Vidicon is cheaper and has a longer operating life, but requires more light. Table 6 lists the key characteristics and costs of both tube types.

Table 5

REPRESENTATIVE STUDIO EQUIPMENT COSTS

Audiotronics		RCA		
"Micropak"	"Maxipak"	"Economy Monochrome TV Studio Package"	"Color-Convertible Monochrome TV Studio Package"	"Medium Color TV Studio Package"
1 monochrome video camera	2 monochrome video cameras, with view finders and 2 zoom lenses	2 monochrome video cameras	2 monochrome video cameras, color convertible	2 color video cameras
1 video tape recorder, monochrome	1 video tape recorder, monochrome, with electronic editing	1 film system, monochrome	1 film system, monochrome	1 color film system
1 video monitor, 16-inch	1 video switcher	1 video tape recorder, monochrome	1 video tape recorder, monochrome	2 video tape recorders, color
	1 audio mixer	1 video switcher	1 video switcher	1 video switcher
	1 control console with program and preview monitors, distribution ampl.	1 audio mixer	1 audio mixer	1 audio mixer
		1 control console	1 control console	1 control console
Cables, mike, tripod	Cables, 2 mikes, tripod, dolly, pedestal	Cables, tripods, mikes, stands, etc.	Cables, tripods, mikes, stands, etc.	Cables, tripods, mikes, stands, etc.
$2,000	$6,500	$21,000	$37,000	$105,000

Table 6

RELATIVE ADVANTAGES OF IMAGE ORTHICON AND VIDICON CAMERA TUBES

Item	Image Orthicon	Vidicon
Cost	High: $1,000–2,000 per tube $15,000–25,000 per camera (b & w) Up to $75,000 for professional quality color camera	Moderate: $200–700 per tube $500–10,000 per camera (b & w) Up to $20,000 for professional quality color camera
Tube life	About 1500 hours	About 5000 hours
Required light levels	Good for low to moderate light levels; can be damaged at high levels	Good only at high light levels; will "smear" otherwise
Operating convenience	Requires more experience to operate (e.g., prolonged focus on stationary image may "burn" tube)	Easier to operate; more rugged
Major applications	1. Highest quality image production 2. Low-light-level use	1. General purpose use, especially for low-budget production 2. Use in film chains where light level is high

Color capability costs several times as much as black-and-white for both Orthicon and Vidicon tubes. Many of the camera components must be triplicated for the three primary colors, resulting in a much more complex and expensive unit. A color camera incorporates a precision optical system that splits the lens image into three parallel primary color beams (red, blue, and green). Three separate pickup tubes are used, each designed for peak efficiency over its own primary color spectrum, and each with its own scanning circuitry. This redundancy, together with the necessity to maintain the correct balance and proportion of the three color signals, is the reason for the great difference in camera cost.

Video tape recorders (VTRs) are becoming increasingly important for cablecasting, since they provide the only practical method to date of storing and playing back video information without conversion to and from another medium. Film, for example, requires optical-to-electronic conversion. The VTR was introduced to the TV broadcasting industry in 1956. Since then, equipment in a wide range of sizes, capabilities, and prices has become available.

The professional standard for broadcasting is the "quadruplex" recorder, using 2-inch-wide tape. (The term "quadruplex" derives from the four magnetic heads used.) Quadruplex recorders are expensive, from $20,000 each for monochrome to more than $100,000 for a high-quality color unit. They also are large and bulky, not suitable for mobile or portable use.

More recently, a second technique called "helical-scan" or "slant-track" recording has been developed, permitting manufacture of smaller, lower-cost VTRs. Helical-scan recorders now come in a variety of tape widths (from 1/2 inch to 2 inches), tape speeds, and performance characteristics. They range in cost from below $1000 for 1/2-inch monochrome units to about $15,000 for 1-inch color recorders. The 1/2-inch units have become particularly popular for outside-the-studio recording since they are portable as well as cheap. Figure 9 illustrates this equipment in use in a series of video workshops for the deaf held recently in Reading and York, Pennsylvania by the Community Video Workshop of Berks-Suburban TV Cable Company.

A major problem with helical-scan recorders has been their lack of standardization. Until recently, tapes recorded on one manufacturer's model could not be played back on another manufacturer's unit. Sometimes different recorders even of the same make and model were incompatible. The result has been to inflict severe limits on program interchange. In New York City, many 1/2-inch tapes prepared by individuals and groups for use on the public access channels must be converted to a different format before they can be played.

Within the past two years, Japanese VTR manufacturers have established standards for 1/2-inch and 3/4-inch equipment that should greatly reduce the compatibility problem. Consequently, cable systems and public-access groups can plan to exchange videotapes with some confidence that they will be playable on other systems.

New cassette and cartridge VTRs promise to expand the use of videotape even further. The 3/4-inch equipment pioneered by Sony and Panasonic produces color pictures of much better quality than the black-and-white tapes made on 1/2-inch machines. And the elimination of reel-to-reel threading makes cassette VTR's easier to use by unskilled operators. From the cable system operator's point of view, cassettes permit greater automation in cablecasting. Presently made cassette equipment is heavier and less portable than 1/2-inch VTRs, however.

MOBILE FACILITIES

As in broadcasting, many events that are of interest for cablecasting do not take place in the studio. The ubiquitous city council meetings and local high school basketball games, so often mentioned as examples of the benefits of cablecasting, are but two cases in point.

The mobile equipment package for covering remote events does not differ in principle from studio equipment. At a minimum, it comprises a video camera, VTR, and a control unit, plus the associated tripod, microphones, and cabling. The control unit contains a video monitor and controls to compensate for local lighting conditions. Power is usually obtained from the vehicle battery or from portable battery packs. Several manufacturers offer mobile packages, from small units that can fit into the back of a passenger car to custom mobile truck installations. The prices vary accordingly, from about $5000 to $30,000 for black-and-white, and proportionately higher for color.

In most cases, the program is recorded on videotape, which is then transported physically to the cablecasting studio for later playback. If real-time cablecasting is desired, a communication link to the studio is necessary. Special short-range, portable microwave links are available for this purpose.

Courtesy: Community Video Workshop Berks-Suburban TV Cable Company

Fig. 9—Using a portable, half-inch videotape recorder and camera

PROGRAM PRODUCTION

The costs of cablecasting hardware (e.g., cameras and VTRs) and technical operating personnel (e.g., cameramen and console operators) can be estimated fairly accurately. Program production and other software costs are more difficult to forecast.

The simplest kind of cablecasting, such as a one-camera recording of a local meeting without editing, may cost as little as $50 to $100 an hour. An hour of blank, 1/2-inch videotape itself costs $20 to $30. Taping a classroom lecture with some editing may run $500 to $1000, although colleges today typically budget $2000 to $3000 per final program hour.[2]

On the other hand, cablecasting an original drama, involving script, cast, costumes, sets, lighting, and rehearsals can easily run into tens or hundreds of thousands of dollars. A cost of $1000 to $2000 per viewing minute is considered a realistic guide for the production of commercial filmed or taped programs with no "star" actors.

Past cablecasting has naturally clustered at the low end of the scale in cost and production quality. Still, local programming has improved steadily from the days of presenting a clock face or weather dial to subscribers. Certainly, the technology is available to support cablecasting as a unique medium in its own right, not as simply a lower-cost, lower-quality imitation of broadcast television.

[2] R. Bretz, *Three Models for Home-Based Instructional Systems Using Television*, The Rand Corporation, R-1089-USOE/MF, October 1972.

IV. INTERCONNECTION OF CABLE SYSTEMS

FORMS OF SYSTEM INTERCONNECTION

In the past, cable systems have served their own communities alone. With very few exceptions, they have not been linked together to exchange programming. However, several forms of interconnection will be important for new cable systems built in the major markets.

Internal Interconnection (Subdistricting)

Neighborhoods or communities within a franchise area may want different programming, owing to ethnic, cultural, or economic interests. Providing certain channels to some areas and not others is called internal interconnection or "subdistricting." For example, New York City requires Manhattan franchises to divide their cable systems into at least 10 subdistricts. Subdistricts presumably would be selected to conform to traditional, identifiable communities, or to political district boundaries.

Area Interconnection

Cities that franchise more than one cable operator often will require that certain programs be shown simultaneously on all systems. Moreover, most large cities will need more than one headend, since present amplifier technology limits the area that can be served from a single point. Consequently, distributing programming simultaneously throughout the city will require interconnection of the multiple headends or hubs, as shown in Fig. 10. Hub interconnection may be particularly important for distributing public access and other community programming to geographically separated neighborhoods that share similar interests. For example, access programming produced in Harlem may be of far more interest to citizens in Bedford Stuyvesant—across the East River and in a different Borough—than to adjacent neighborhoods on Manhattan's upper east side.

Area interconnection may also extend beyond a single jurisdiction, especially to include a city and its surrounding suburbs. Here there are several factors to be considered. Metropolitan-wide interconnection may be important in strengthening common bonds between city residents and suburbanites, rather than creating elec-

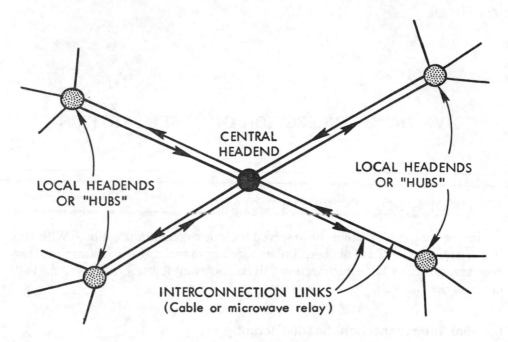

Fig. 10—Area interconnection of multiple hubs

tronic barriers between them. New services, moreover, such as televised university extension classes and sports events on pay TV, may not be feasible unless distributed simultaneously to a large base of subscribers in a number of cable systems. Finally, suburban cable systems in the major markets may not prosper unless they are interconnected with other suburbs or the central city. However, metropolitan-wide interconnection also raises issues of possible cross-subsidy and equitable sharing of interconnection costs. These points are discussed in recent studies of metropolitan cable systems.[1]

Networking

Networking implies interconnection of cable systems that are not physically adjacent, primarily to receive the same programming simultaneously. It offers a strong economic incentive for program producers and advertisers by increasing the audience above that reached by a single system. Networking is exactly the opposite of community-oriented programming, of course, both in philosophy and in the interconnection technology involved. Nevertheless, a cable system may have to provide both kinds of capability.

Cable systems are now thinking of networking on a regional or national basis, probably by satellite. Several companies have proposed to distribute programming

---

[1] Cable Television Information Center, *Cable Television: Options for Jacksonville*, The Urban Institute, Washington, D.C., February 1973; and Leland L. Johnson, et al., *Cable Communications in the Dayton Miami Valley: Basic Report*, The Rand Corporation, R-943-KF/FF, January 1972.

to cable systems when the first U.S. domestic satellite systems become operational after 1975. National networking on a trial basis using the Canadian satellite system may begin this year.

INTERCONNECTION TECHNIQUES

Internal interconnection requires a separate trunk line for each subdistrict and switching equipment at the headend. Switching costs are minor if only one or two channels are involved—for example, switching a single public access channel at the headend to carry programs originated in one of the subdistricts. It will cost much more to enable each subdistrict to receive access programs simultaneously from all the other subdistricts.

Headend or hub interconnection will be by cable or microwave relay. The basic problem is to degrade signals as little as possible between headends, but some loss of signal quality is inevitable. Consequently, system designers must take interconnection losses into account in laying out the distribution system around each hub. In the worst case, signals travel from the far end of one system to its headend, then via the interconnection path to another headend, and finally downstream to the far end of the second system. The signals will go through twice as many amplifiers and hence suffer twice as much degradation as signals that are received off-the-air at a headend.

Cable interconnection uses a large-diameter coaxial cable, often called "supertrunk," to reduce losses. Interconnected signals may also be translated to lower frequencies where attenuation is less. The larger the cable and the fewer the channels carried, the greater the distances between headends that can be connected with supertrunk. Supertrunk connecting hubs five to ten miles apart may be able to carry 20 channels or more, if each hub distribution system has been well designed. Supertrunk runs can be 15 miles or longer if only a few channels are carried. A separate cable must be installed between each pair of interconnected points, of course, and the total cost is proportional to the total number of cable miles required. Two cables may be needed for each link if signals are to be transmitted in both directions.

Microwave relay often has signal-quality and cost advantages over supertrunk when many points must be interconnected, or when hubs are more than 10 miles apart. All microwave links must be licensed by the FCC. The FCC has allocated a 250 MHz bandwidth in the microwave region for "Cable Television Relay Service," or "CARS" (an acronym from its former designation, Community Antenna Relay Service). Single-channel CARS band transmitters and receivers originally were developed to relay distant signals in a series of "hops" of 25 to 30 miles. No more than 10 television channels can be carried from one point to another with this equipment.

Since urban cable systems may need interconnection of more than 10 channels, manufacturers recently have developed equipment to transmit up to 18 channels from a central point to outlying hubs. Microwave links of this sort are known as Local Distribution Service (LDS). The central LDS transmitter is more expensive than a single-channel CARS band transmitter, but the overall system cost may be much lower if many interconnection channels are required. Consequently, LDS

Table 7

COMPARISON OF CABLE AND MICROWAVE INTERCONNECTION FEATURES

Feature	Cable	Microwave (CARS)	
		Single-Channel FM	Multichannel AM or FM "Local Distribution Service"
Channel capacity	20-30 channels per cable	Limited to 10 channels by FCC restrictions	2 groups, each with 18 operational and 1 calibration channel; normally only one group is assigned in each area
Capital cost	About $5000-7000 per mile (aerial installation)	$7500-15,000 per channel	About $100,000 for an 18-channel transmitter; $5000-6000 for each receiver
Reliability	Excellent (assuming maximum trunk cable distances not exceeded)	Good (subject to some atmospheric disturbance)	Subject to some atmospheric disturbance; long-term reliability not established due to newness of equipment
Maximum transmission distance for good quality signals	5-15 miles, depending on cable diameter and bandwidth to be carried	20-35 miles	15-25 miles
Two-way capability?	Yes, but generally requires two cables per link at about a 50-percent increase in cost per mile	Yes, at a cost increase of about $5000-10,000 per reverse channel	Not with present systems

promises to be a popular technique for high-capacity areawide interconnection. However, present LDS systems are one-way only; returning signals from outlying hubs to the central point, or transmitting signals among headends, would require separate cable or microwave links.

The choice of technique thus involves the number of interconnected headends, the number of channels carried, distances, and whether one-way or two-way transmission is required. Table 7 gives a brief comparison between cable and microwave interconnection. Some systems may use a combination of the two—for example, LDS for multichannel transmission from a master headend to outlying hubs, with cable connections back to the master headend for return communications. The best area interconnection technique must be worked out in each individual case.

V. TWO-WAY COMMUNICATIONS

Besides transmitting "downstream" to the home, a coaxial cable can carry information back "upstream" from subscribers. This ability suggests a great variety of new services and gives cable television much of its "blue-sky" allure.[1]

The FCC now requires each new <u>major-market cable system</u> to "<u>maintain a plant having technical capacity for nonvoice return communications</u>." This is interpreted to mean that the system meets the FCC's intent if it *eventually* can provide return communication (from subscribers to the headend) without "time-consuming and costly system rebuilding."[2]

A listing of proposed two-way or interactive services, adapted from a previous Rand report, appears as Table 8. These can be categorized broadly into two groups: services for individual subscribers, and services for institutions.

SUBSCRIBER SERVICES

Television pictures, voice conversations, and data messages are all carried on the cable as electrical signals; they differ primarily in the frequency bandwidth and the subscriber equipment each requires. In principle, a subscriber could send data, voice, or video signals upstream to the headend or to other subscribers on a two-way cable system. In practice, bandwidth constraints and terminal costs will make televi-

---

[1] For additional discussion of two-way cable communications, see Ward, op. cit.; Walter S. Baer, *Interactive Television: Prospects for Two-Way Services on Cable,* The Rand Corporation, R-888-MF, November 1971; Ronald K. Jurgen, "Two-Way Applications for Cable Television Systems in the '70s," *IEEE Spectrum,* November 1971, pp. 39-54; W. F. Mason, et al., *Urban Cable Systems,* The MITRE Corporation, M72-57, May 1972; Hubert J. Schlafly, "The Computer in the Living Room," TelePrompTer Corporation, New York, 1973; and Gerald M. Walker, "Special Report: Cable's Path to the Wired City is Tangled," *Electronics,* May 8, 1972, pp. 91-99.

[2] Sec. 129 of the FCC *Cable Television Report and Order,* 37 Fed. Reg. 3252, 1972 reads as follows:

129. We are not now requiring cable systems to install necessary return communication devices at each subscriber terminal. Such a requirement is premature in this early stage of cable's evolution. It will be sufficient for now that each cable system be constructed with the potential of eventually providing return communication without having to engage in time-consuming and costly system rebuilding. This requirement will be met if a new system is constructed either with the necessary auxiliary equipment (amplifiers and passive devices) or with equipment that could easily be altered to provide return service. When offered, activation of the return service must always be at the subscriber's option.

37

Table 8

SOME PROPOSED INTERACTIVE SERVICES FOR CABLE TELEVISION[a]

Subscriber	Institutional
Interactive instructional programs	Computer data exchange
Fire and burglar alarm monitoring	Teleconferencing
Television ratings	Surveillance of public areas
Utility meter readings	Fire detection
Control of utility services	Pollution monitoring
Opinion polling	Traffic control
Market research surveys	Fingerprint and photograph
Interactive TV games	identification
Quiz shows	Civil defense communications
Pay TV	Area transmitters/receivers for
Special interest group conversations	mobile radio
Electronic mail delivery	Classroom instructional TV
Electronic delivery of newspapers	Education extension classes
and periodicals	Televising municipal meetings
Remote calculating and computer	and hearings
time sharing	Direct response on local issues
Catalog displays	Automatic vehicle identification
Stock market quotations	Community relations programming
Transportation schedules	Information retrieval services
Reservation services, ticket sales	Education for the handicapped
Banking services	Drug and alcohol abuse programs
Inquiries from various directories	Health care, safety, and other
Local auction sales and swap shops	public information programs
Electronic voting	Business transactions
Subscriber originated programming	Credit checks
Interactive vocational counseling	Signature and photo identification
Local ombudsman	Facsimile services
Employment, health care, housing,	Industrial security
welfare, and other social	Production monitoring
service information	Industrial training
Library reference and other	Corporate news ticker
information retrieval services	Telediagnosis
Dial-up video and audio libraries	Medical record exchange
Videophone	

[a]It is unlikely that all of these services will be economically feasible on cable television networks. Some may not even be socially desirable. They have been compiled from various reports, FCC filings, corporate brochures, and advertising materials. Adapted from Baer, *Interactive Television.*

sion origination from the home too expensive for all but specialized applications.[3] Eventual video links among subscribers are much more likely to come about on the switched telephone network than on a party-line cable system. The telephone also will remain a better choice for private conversations.[4]

Consequently, an upstream return path from the home will be used principally to carry data messages from subscribers. These messages could include responses to questions asked on an educational program, opinions on a proposed city ordinance, and orders to buy the sewing machine just advertised on the screen. Fire and burglar alarm messages also could be sent automatically to a central station.

These messages all have common characteristics. They require much less bandwidth than voice conversations, and they can be encoded in digital form for rapid computer processing. Moreover, digital messages from thousands of subscribers can be packed together into a single data stream that uses the upstream cable capacity very efficiently.

Each subscriber would have his own digital code or "address" for two-way response services. A computer at the headend (or some other suitable location) would query each subscriber in turn, using a special downstream channel. This technique is known as "polling." A two-way terminal attached to the cable and TV set would record the subscriber's messages, store them, and send them upstream when the terminal was polled. The messages would then be recorded at the headend computer or sent on to the city council chambers, the police station, or the department store.

The basic subscriber response terminal looks like a small box with a telephone-like keyboard and a lock to prevent unauthorized use (Fig. 11). A tunable converter might be built in, and smoke sensors, burglar alarms, and utility meters could be connected to it. Several companies are now experimenting with prototype subscriber terminals. They cost close to $1000 today, but industry sources estimate that further development and mass production may reduce the price to $100 or so by 1980.

Some specific two-way services such a terminal can handle include:

- Counting the number of subscribers tuned to a given channel
- Ordering an air conditioner advertised on "special sale"
- Requesting a pay-TV movie
- Answering a multiple choice quiz presented as part of a televised college class
- Responding to a political opinion poll
- Reading utility meters automatically
- Monitoring fire and burglar alarms

Other services, such as browsing through a catalog displayed on the TV screen, making theater reservations, or requesting a paragraph from the *Encyclopaedia Britannica* will require more complex subscriber terminal equipment.

---

[3] For example, two-way video links between a classroom and handicapped children at home were demonstrated on the cable system in Overland Park, Kansas. Continuing service was not economically feasible, however, and the demonstration was terminated after a few months.

[4] Cable systems might install a shared upstream voice channel, although cable's advantage over the telephone seems questionable. For example, a city council meeting televised on a two-way cable system might allot time for audience questions and comments. With the proper equipment at home and in the city council chambers, a subscriber could signal to be heard and then wait his turn until an upstream voice channel were available. While his question was being transmitted upstream, that voice channel would be denied to other users.

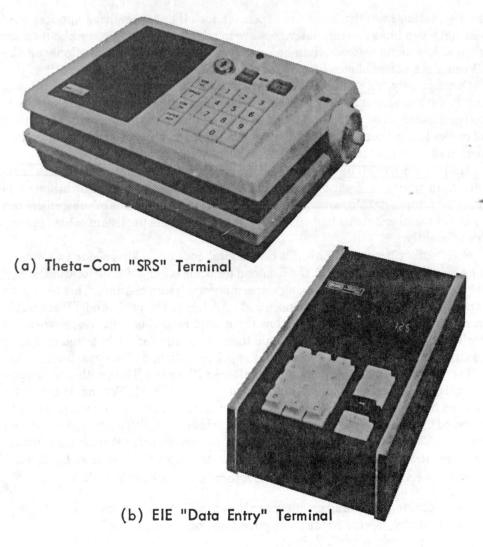

(a) Theta-Com "SRS" Terminal

(b) EIE "Data Entry" Terminal

Fig. 11—Typical terminals for two-way subscriber response services

INSTITUTIONAL SERVICES

Institutional users of the cable system may want and be able to afford more expensive terminals and greater upstream capacity. These users include:

- Businesses that want high-speed computer data exchange
- Industrial plants that televise extension courses with student feedback to a nearby university
- Schools that want two-way video for in-class instruction and after-school teacher meetings
- Hospitals that exchange medical records and diagnostic test results, and hold two-way video consultations
- Police agencies that transmit fingerprints and photos among precinct houses, or monitor streets and public areas with remote TV cameras
- Local government agencies that want two-way video links for teleconferencing

Experiments in each of these uses are under way today, although they generally use microwave links, closed-circuit systems, or the telephone network rather than cable.[5]

TWO-WAY TRANSMISSION TECHNIQUES

Two-way services by definition require a cable system that can transmit information in both directions. The chief technical difficulty today is that each TV set or subscriber terminal introduces some noise into the upstream transmission path, and the cumulative effect from large numbers of terminals may be intolerable. Better receivers, taps, filters, and other components must be designed before two-way cable systems can accommodate the tens of thousands of subscribers contemplated in major markets.

The two basic technical approaches to two-way transmission are:

1. Use separate cables for upstream and downstream transmission;
2. Send signals in both directions simultaneously on the same cable, using different frequency bands to separate the upstream and downstream signals.

A third approach would use a one-way round-robin cable loop to bring signals to and from subscriber locations. While interesting conceptually, this technique has not yet been applied on operating cable systems. The three approaches are shown schematically in Fig. 12.

Having a separate cable for upstream transmission presents fewer technical problems and offers more upstream capacity, but is more expensive. An ordinary telephone wire or wire pair used for the upstream path would satisfy the FCC's

---

[5] Further discussion of municipal and educational applications can be found in Robert K. Yin, "Applications for Municipal Services," and Polly Carpenter, "Uses in Education," Chapters 2 and 3, respectively, of Polly Carpenter et al., *Cable Television: Developing Community Services*, Crane, Russak & Co., New York, 1974.

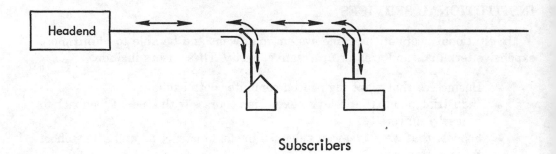

(a) Two-way transmission on a single cable

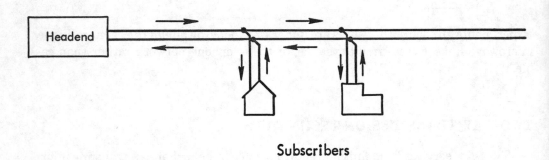

(b) Separate cables for upstream and downstream transmission

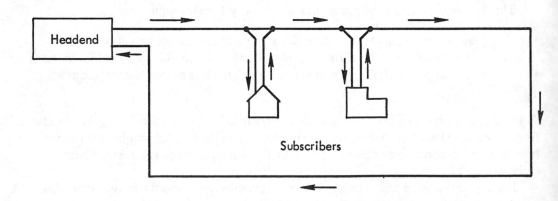

(c) Round-robin cable loop

Fig. 12—Three techniques for two-way transmission

requirement for "nonvoice return communications," but would provide little capacity for future services.

Carrying signals in both directions simultaneously on a single cable costs less than installing separate cables but is more complex. Upstream and downstream signals must be separated in frequency to avoid interference. The most popular approach today is to use the sub-VHF bandwidth below channel 2 (below 54 MHz) for upstream signals, retaining the 54 to 300 MHz bandwidth for downstream transmission. This is known as the "subsplit" or "low-split" technique.

The coaxial cable itself is bidirectional and poses no problems to two-way transmission. Amplifiers, however, are one-way devices and require some by-pass path for signals going in the opposite direction. Consequently, the upstream and downstream signals must be routed to different amplifiers and must be separated by electronic filters, as shown in Fig. 13. In practice, filters do not separate frequencies as sharply as might be desired. To allow enough of a "guard-band" to prevent interference, most system designers restrict upstream signals to no higher than 30 MHz, rather than 54 MHz. A similar precaution is taken not to use frequencies below about 5 MHz, since the electronic filters used in amplifier power supply circuits cut off these low frequencies. As a result, the subsplit approach limits upstream transmission to about 5 to 30 MHz, a 25 MHz bandwidth equivalent to four standard TV channels.

The single-cable approach is popular with cable operators because its installation cost is lower. Moreover, if the system has no immediate plans to use upstream communications, only the downstream amplifiers need be installed initially. Purchase and installation of the filters and upstream amplifiers shown in Fig. 13 can be deferred to a later date. Amplifier manufacturers offer models with housings large enough to accommodate plug-in modules for activating upstream communications in the future. At present, this satisfies the FCC requirement for "technical capacity for nonvoice return communications."

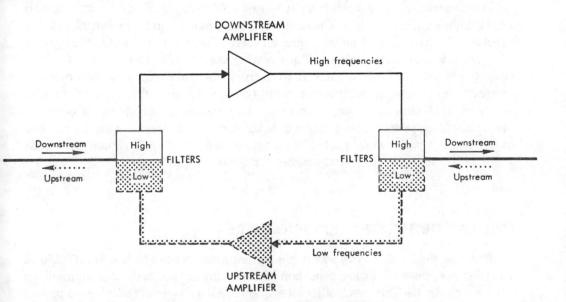

Fig. 13—Simultaneous two-way communication on a single cable

Although economically attractive, the subsplit technique suffers first from its limitation in upstream bandwidth. Although 25 MHz may be sufficient for subscriber response data services, heavy institutional demand for two-way video channels will require additional upstream capacity. As a result, a "midsplit" technique has been devised that divides upstream and downstream signals between 108 and 160 MHz, permitting 14 or more 6 MHz channels in each direction. The subsplit and midsplit techniques are compared in Fig. 14.

Of course, a single cable system built in a major market could not adopt the midsplit technique, since it then could not deliver 20 downstream channels as the FCC rules require. A midsplit system thus implies two or more cables—the first carrying 20 or more channels downstream, and the second with the capability for 14 or more channels in both directions. This second cable can exist in "shadow" form, however, as described in Sec. II, with only the cable and housings for electronics installed initially, and the electronic components added later. The single operating cable plus shadow-trunk design is currently very popular with operators, since it entails (initially) only the extra cost of the second cable and provides both one-way and two-way expansion capacity for the future.

Still, the technical problems have not been fully resolved, particularly for the subsplit approach. Designing filters for two-way operations is difficult; each filter will add some distortion to the downstream video signal, and the cumulative effect of many amplifiers in cascade may be an unacceptable picture. Although the problem should be worked out in time through experience and better filter design, it has been a barrier to early two-way development. At the very least, the need for two-way filters will restrict the number of amplifiers that can be cascaded, and thus further limit the area that can be served from a given headend.

In summary, a simplified cost comparison of several two-way system designs is shown in Table 9. These estimates should be compared with the costs of one-way systems listed in Table 4.

The separate-cable approach to two-way cable communications is presently more reliable, but more costly. The single-cable approach minimizes initial cost, but is technically more risky. Neither approach solves the problem of additive upstream noise from subscriber terminals. Once again, the benefits from added two-way capacity are more difficult to quantify than the costs, and there is no "best answer" for every community. Some cities may want to require dual trunk and feeder cables —at least in shadow form—that can be used to expand either upstream or downstream capacity, or some combination of both. A city that foresees a need for extensive two-way capacity among schools, hospitals, and other institutional users may prefer to install entirely separate cables for these applications.

TWO-WAY DEMONSTRATION PROJECTS

The first field tests of two-way cable communications began in 1971. (Table 10 lists the major ones.) They are being conducted today only as tests of equipment for subscriber response services, not as attempts to deliver two-way services to paying subscribers. As of May 1, 1973, fewer than 100 two-way subscriber terminals were in active operation in the United States. Additional tests are planned in the United

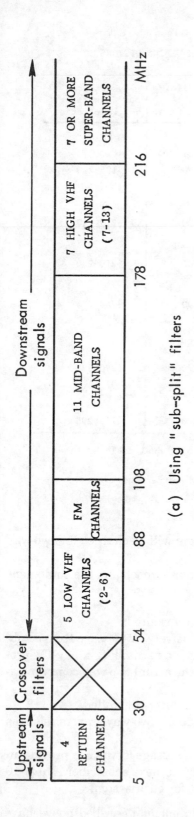

(a) Using "sub-split" filters

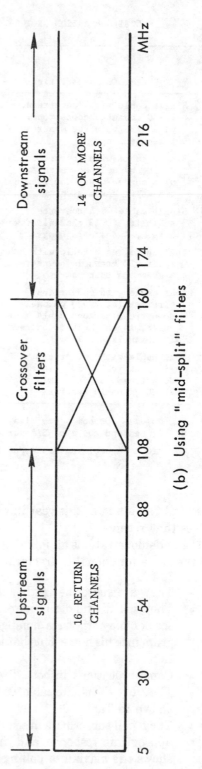

(b) Using "mid-split" filters

Fig. 14—Subsplit and midsplit techniques

Table 9

COST COMPARISON OF TWO-WAY CABLE SYSTEM DESIGN OPTIONS

System Characteristics	Cost per Mile, Aerial Cable Distribution Plant	System Cost per Subscriber, 40% Penetration[a]
Single cable with converters, 20–30 channels downstream, plus wire pair for upstream transmission	$5800–6800	$195–220
Single cable with converters, 20–30 channels downstream, plus shadow trunk for eventual two-way transmission	$6500–7500	$210–240
Single cable with converters, subsplit, 20–30 channels downstream, 4 channels upstream	$6500–8000	$210–250
Dual cable, each cable with subsplit, 12 channels downstream,[b] 4 channels upstream	$8800–9500	$240–255
Dual cable with converters: first cable, 20–30 channels downstream; second cable midsplit, 14 channels downstream, 16 channels upstream	$9500–11,000	$290–330
Dual cable with converters: first cable, 20–30 channels downstream; second cable, 20–30 channels upstream	$7800–9800	$245–295

[a]Assumes 100 homes per mile, converters at $35, dual cable switches at $4, headend costs of $15 per subscriber.

[b]Minus channels unusable due to on-channel interference.

States and Canada later this year in which subscribers will be charged for two-way services they receive.

These demonstrations all follow the same general approach and use equipment with the same functional characteristics; that is:

- They are subscriber rather than institution oriented.
- They use digital data communications exclusively. A message from a subscriber may trigger a nondigital action, such as switching a pay-TV program into his home, but that is ancillary to the actual two-way communications.
- Communications in both directions are computer-controlled.
- They use the single-cable, subsplit technique for two-way transmission, as shown in Table 10.
- They poll subscribers in sequence to see if a message is waiting, provide an appropriate response, and then proceed to the next subscriber. Table 10 shows the maximum polling rate designed for each system.

To no one's surprise, a number of the technical problems described above have shown up in these tests. Additive upstream noise has been a particularly vexing

Table 10

CURRENT FIELD TESTS OF TWO-WAY CABLE SYSTEMS

Equipment Manufacturer	Test Location	Transmission Frequency Band (Mhz)		Subscriber Polling Rate per Second (Design Limit)	Status
		Downstream	Upstream		
Theta-Com	El Segundo, Calif.	108-112	21-25	5000	20-25 terminals installed in first phase; plans for 1000 terminals within one year
Electronic Industrial Engineering	Orlando, Fla.	88-108	12-14	1000	24-27 terminals installed in first phase; plans for 500 terminals in 1973
Scientific-Atlanta	Carpentersville, Ill.	111.1-113.9	0.225-0.275 or 5.25-5.75	820	6 terminals installed for fire and burglar alarms; plans for operations later in 1973
TOCOM	Irving, Tex.	50	6-30	1000	Planned for installation later in 1973

problem. Still, the technology for two-way subscriber response services is known, and the engineering problems will be resolved in time. By 1980 most large U.S. cable systems should have operating two-way transmission facilities. The prospects for using these facilities for two-way subscriber or institutional services are less clear, however, and await the first full-scale market tests sometime later in this decade.[6]

[6] The Federal Government may also play a role in initiating two-way demonstration projects. In 1972, the President's Office of Telecommunications Policy released a major study and requested statements of qualifications from organizations that might plan demonstration projects using telecommunications technology. See *Pilot Projects for the Broadband Communications Distribution System*, Malarkey, Taylor & Associates, Washington, D.C., November 1971; and *Commerce Business Daily*, August 25, 1972, p. 12, for details. However, no federal support for such projects has yet been announced.

VI. TECHNOLOGY FOR NEW SERVICES

PAY TV AND PRIVATE CHANNELS

Pay television (or pay-cable, as the FCC now terms it) will be the first new service offered by most cable systems. Viewers will pay an extra charge (over and above the basic cable subscription fee) to watch first-run movies, sports events, theater productions, and other special programs. Pay-TV movies have already proved successful in hotels and motels. Several Pay-TV experiments now are under way on cable systems (Table 11), and cable operators estimate that they will have more than one million pay-TV subscribers by the end of 1974.

Subscribers will watch pay-TV programs on their standard TV sets. New equipment will be needed, however, to deny the programs to those who do not want to pay for them, and to record the proper charges for those who do.

The simplest technical approach is to send pay-TV programs on special frequencies—perhaps in the midband or superband—and furnish pay-TV subscribers with a special converter to receive them. To guard against subscribers tapping in for free, pay-TV promoters usually code or scramble the signals at the headend. A decoder as well as a converter is then required in the subscriber's pay-TV terminal. Some of the first pay-TV terminals are illustrated in Fig. 15. For further security the operator can install a switch inside the terminal that can be turned on and off only from the headend. He might even take the decoding terminal out of the home entirely and place it at the tap or at a feeder amplifier. Each step is successively more costly, but provides greater protection. The same cost and security considerations would apply to private channels that only certain subscribers are entitled to receive. Private channels might be needed for police training classes, for example, or for assuring privacy for banking transactions in a two-way system.

The simplest form of billing is to charge a fixed amount each month, no matter how much or how little pay-TV programming the subscriber watches. Both pay-TV promoters and viewers prefer, however, that charges be made on a per-program basis. To do this, the subscriber must be able to signal when he wants a particular program and to be billed accordingly.

In one currently demonstrated pay-TV system, subscribers purchase a "ticket" —shaped somewhat like a plastic credit card—in advance to watch particular pay-TV programs. When the program is shown, he places the ticket in a special slot in his pay-TV terminal. The ticket contains magnetic strips or holes that activate decoder circuits to provide an undistorted picture on the screen. Each program is

Table 11

PAY-TV EXPERIMENTS ON CABLE TELEVISION SYSTEMS

City	Cable Operator	Pay-TV Operator	Type of System	Beginning Date	Programs Offered
Wilkes-Barre, Allentown, other Pa. cities	Service Electric	Home Box Office (partially owned by Time, Inc.)	one-way, monthly fee	Nov 1972	Sports, movies
Sarasota, Fla.	Storer Broadcasting	Theatre Vision	one-way, with tickets	Jan 1973	Sports, movies, cultural events
Reston, Va. Olean, N.Y. Clearfield, Pa. Pottsville, Pa.	Warner Communications	Gridtronics (subsidiary of Warner Communications)	one-way, monthly fee	Feb 1973	Movies
San Diego, Calif.	Cox Cable	Optical Systems	one-way, with tickets	Mar 1973	Sports, movies
Long Beach, Calif.	Times-Mirror CATV	Cinca Communications	one-way, monthly fee	Mar 1973	Movies

NOTE: These pay-TV experiments were actually under way on cable systems as of April 1, 1973. Additional experiments are planned for 1973 or early 1974 by Cablecom-General/Home Theatre Network, Trans-World Communication, TelePrompTer/Magnavox, American Television & Communications/RCA, Theta Cable/Theta-Com, and others.

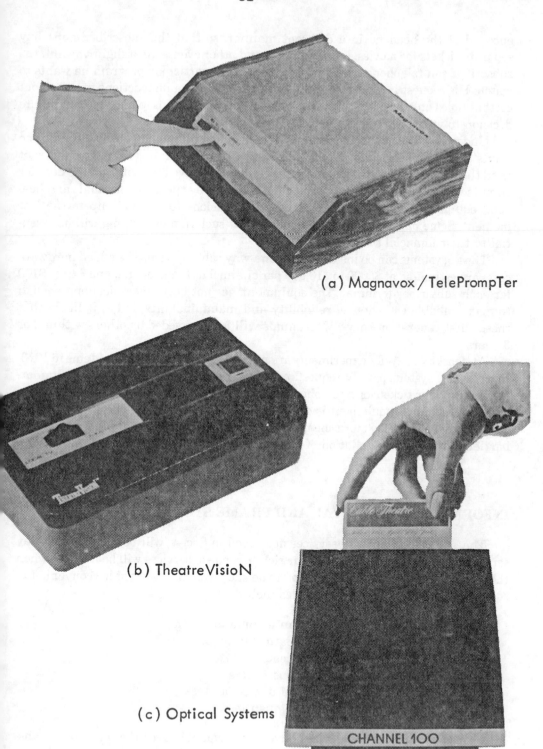

(a) Magnavox/TelePrompTer

(b) TheatreVisioN

(c) Optical Systems

CHANNEL 100

Fig. 15—Home terminals for pay television

encoded at the headend in a different manner, so that the subscriber must buy separate tickets for movies, hockey games, and other events. As a slight variant, the subscriber can telephone to find out the code for a particular program he wants to watch. His request is recorded for billing, and he is given instructions to punch out certain holes on a previously unused ticket. This allows him to make a last-minute decision to watch a pay-TV program.

Under a different scheme, the subscriber calls in for each pay-TV program he wants to see. No previously purchased tickets are required. Instead, a signal is sent from the headend when the subscriber's request is received to switch on the decoder circuits in his terminal. This approach requires more expensive terminals and headend equipment, but it permits all pay-TV decisions to be made up to the last moment. Some pay-TV promoters consider that such impulse buying will be essential to their financial success.

These systems can be installed on a one-way cable system. The added hardware costs are estimated at $5000 to $10,000 per channel at the headend, and $40 to $100 for each subscriber terminal. The equipment has not been in service long enough to make reliable estimates of reliability and maintainability, but it is likely that these "first generation" pay-TV terminals will be superseded by others within 2 or 3 years.

At least two pay-TV experiments are planned on two-way cable systems in 1973. In a two-way system, pay-TV requests could be sent upstream on the cable, eliminating the need for tickets or a telephone return link. Two-way pay-TV also has other advantages, but its equipment is considerably more expensive. The very fact that pay-TV can be installed more cheaply as a one-way service may prove a formidable barrier to the early installation of two-way home terminals.

INFORMATION RETRIEVAL AND FRAME STORAGE

The subscriber response services described in Sec. V will be the first two-way services developed for the home. A far richer range of services will be possible when the subscriber can request information to be shown on his TV set at his convenience. A few of the many potential examples include:

- Getting help in filling out an income tax form
- Buying theater tickets for Saturday night
- Balancing one's checking account
- Seeing a map of alternate bus routes
- Finding the last sale price of a specific stock
- Taking a computer-aided Spanish lesson

In the past, some of these services have been available over the telephone. Cable now provides a new way to transmit information and have it displayed on the TV screen.

Information retrieval services will be considerably more complex and costly than subscriber response services. They require much more sophisticated computer and information storage facilities at the headend, and an even greater investment in computer programming or "software." They also demand a more sophisticated subscriber terminal that can display requested information as a still picture on the TV screen. The problem is somewhat like that of projecting different frames of a movie film on different screens, one frame at a time, for different viewers. One subscriber might want to see current football scores; another, airline schedules; a third, children's clothes in a merchandise catalog; and so forth.

Like movies, television presents a series of still pictures ("frames") on the screen rapidly enough that the eye perceives small changes as motion. The U.S. television standard is 60 frames a second. The technology exists for sending these separate frames to different subscribers, with each frame captured at a subscriber's terminal and replayed over and over for viewing on his TV set. Such terminals are generally known as "frame grabbers" or "frame stoppers." Prototype frame stoppers have been developed using videotape recorders or special "storage tubes" to capture and replay single frames. In the approach recently demonstrated in Reston, Virginia, up to 600 subscribers could in principle be served simultaneously over a single 6 MHz television channel, if new frames were requested every 10 seconds on the average. This demonstration used a telephone line for the return link from subscriber to headend, but the Reston cable system is now being retrofitted for two-way transmission.[1]

As might be expected, frame-stopping terminals would be very expensive today, even if two-way cable systems were operational. Although development is proceeding rapidly, frame-stopping terminals are still likely to cost considerably more than a color television set in 1980. Consequently, information retrieval services on cable will be available to business, government, and institutional subscribers well before they are feasible in the home.

ELECTRONIC MAIL AND FACSIMILE

Another class of proposed cable services would deliver letters, newspapers, periodicals, and other documents to the home electronically rather than through the mails. Cable's huge capacity makes it potentially more attractive than the telephone network for large-scale, electronic document delivery.[2] Using digital addresses, a personal letter could be sent to only one subscriber, while advertising flyers were distributed to an entire neighborhood.

Subscriber terminals are again the pacing technical problem. A frame-stopping device could display documents on the television screen a few paragraphs at a time, if the subscriber did not want to keep a printed copy. However, proponents of electronic mail delivery generally assume that some sort of facsimile or "hard copy" device would be needed in the home. It would scan a downstream channel for addressed digital data, separate the information from the high-frequency carrier, and record text and illustrations on a roll of paper. Developing such devices poses

---

[1] K. J. Stetten and R. K. Lay, *A Study of the Technical and Economic Considerations Attendant on the Home Delivery of Instruction and Other Socially-Related Services Via Interactive Cable TV:* Vol. 1, *Introduction and Summary,* The MITRE Corporation, M72-200, December 1972; J. Volk, *The Reston, Virginia Test of the MITRE Corporation's Interactive Television System,* The MITRE Corporation, MTP-352, May 1971.

[2] See Baer, *Interactive Television,* pp. 21-23; and W. B. Gross, "Distribution of Electronic Mail over the Broadband Party-Line Communications Network," *Proceedings of the IEEE,* Vol. 58, No. 7, July 1970.

problems of paper storage and electric power cost, as well as that of building a cheap, reliable home unit. Facsimile devices coupled to cable should be available for institutional use by 1980, but they are even less likely to be used in the home by then than are frame-stopping terminals.

VIDEOCASSETTE RECORDERS AND CABLE TELEVISION

A videocassette recorder can record television programs for playback at a later time. As described in Sec. IV, many cable systems will use videocassette equipment to record and play programs from the headend. The elimination of tape threading and open reels greatly simplifies equipment handling.

Videocassette (or video cartridge) recorders now are also being marketed for home use. They can, of course, record broadcast as well as cable television programs, so that they have no direct connection with new cable services. Still, the combination of cable channel capacity and videocassette storage suggests some interesting possibilities. As examples, adult extension courses or pay-TV programs may be offered on cable at inconvenient times for some viewers. An interested subscriber could connect his videocassette recorder to record these programs when transmitted, and then play them back on his TV set at his convenience. Consequently, the availability of cassette recorders should increase the markets for specialized cable programs and perhaps make some marginal ones economically attractive. Cassette recorders may also be used as frame-stoppers or for electronic document delivery, as described above.

Today, videocassette recorders cost upwards of $1000 and are unstandardized. No clear winner has yet emerged from the competing technologies using videotape, film, or recording discs. Costs are sure to come down in the future, however, so that these devices may become standard household items in the 1980s. Some cable system operators may in fact rent cassette units to subscribers to enhance the market for their services.

HIGH-RESOLUTION TV

Cable systems eventually will deliver bigger, brighter, and sharper television pictures than are standard today. Although cable service is usually purchased to improve reception, television picture quality is limited intrinsically by the channel bandwidth and by the poor reception characteristics of most TV sets. Some observers believe that viewers will demand higher quality sound and pictures before they will pay a separate charge for movies, operas, ballets, and other performances.

If demand warranted, these events could be transmitted over a 10 or 12 MHz channel and displayed on special "high-resolution" TV receivers. High-fidelity, stereo sound could accompany the higher-quality pictures. A well-designed cable system could carry these channels without affecting transmission of the standard 6

MHz channels, whereas it would be very difficult to reallocate the broadcast spectrum to accommodate high-resolution television.[3]

High-resolution TV sets for the home remain to be developed. The scientific principles are well known, but difficult technical and cost barriers remain in the way of producing high-resolution receivers for consumers. Moreover, standards for high-resolution TV would have to be adopted by program producers, cable distributors, and receiver manufacturers before the service could be provided. For these reasons, high-resolution TV in the home is not likely to be widely available before 1990, although it may be feasible in theaters and group viewing centers well before then. Video projection systems that offer larger screen size but no higher resolution are also under development.

INSTITUTIONAL SERVICES

All the above new services are ultimately targeted at mass subscriber audiences, although they may be feasible for businesses, schools, and government agencies before they reach the home. Some cable applications will be developed expressly for institutional users, however. Among others, these include:

- Teleconferencing and other two-way video uses
- Video surveillance
- High-speed data transmission

Teleconferencing

Two-way audio and video links may enable people at scattered locations to hold meetings and discussions. Government agencies, businesses, hospitals, and schools can all use teleconferencing to save time and establish more effective patterns of communication.[4] A consortium of local governments around New York City, the Metropolitan Regional Council, is now developing teleconferencing links by microwave to improve local agency cooperation and administration.[5] Two-way video links between a hospital and outpatient clinics provide a more dramatic example. One such prototype system permits specialists at Massachusetts General Hospital to diagnose patients at the emergency medical station at Boston's Logan Airport.[6] This system again uses a microwave link, but a two-way cable system would offer an attractive alternative if available.

The equipment for teleconferencing is similar to that long used in closed-circuit

[3] E. W. Herald, "A Compatible High-Resolution TV System for Cablecasting," *Proceedings of the IEEE*, Vol. 58, No. 7, July 1970.

[4] Obviously, teleconferencing cannot substitute for all face-to-face meetings and discussions. The tradeoffs are subtle and, for the most part, unknown. See Richard C. Harkness, *Communications Innovations, Urban Form and Travel Demand*, Research Report No. 71-2, Urban Transportation Program, University of Washington, Seattle, January 1972, for a survey and bibliography of research in this area.

[5] D. J. Alesch, *Intergovernmental Communication in the New York-New Jersey-Connecticut Metropolitan Region*, The Rand Corporation, R-977-MRC, May 1972.

[6] For further discussion of this and other health services applications, see Yin, *Cable Television: Applications for Municipal Services*, and Kas Kalba, *Communicable Medicine: Cable Television and Health Services*, report prepared for the Sloan Commission on Cable Communications, September 1971.

television systems. For transmission, a video camera and microphone pick up the desired video and audio signals. Special close-up camera attachments may be needed for medical diagnosis or for viewing a printed page or drawing. The outgoing signals are amplified and modulated (i.e., translated to a frequency band authorized for that service), and then sent out over the cable. A video tape recorder can store programs for later transmission and record incoming signals. The video display generally is on a standard TV receiver or monitor, although some applications may require higher resolution or larger screen size. If standard displays are used, the equipment cost at each teleconferencing station will range from $2000 to $5000 for portable black-and-white components, to $20,000 to $40,000 for full color capability.

Each teleconferencing link requires separate 6 MHz channels for upstream and downstream communication, unless special bandwidth compression schemes are employed.[7] However, several stations can share channels, with only one permitted to transmit at a time. Still, heavy use of two-way video would use up a large share of the cable's capacity or, more likely, require installation of separate cables for teleconferencing.

Video Surveillance

Several cities have experimented with monitoring streets and other public areas by remote video cameras connected to a central point. Traffic and crime control are the two commonest applications.[8] These are essentially one-way (upstream) video transmission services, sometimes requiring a narrowband downstream link for remotely positioning and controlling the cameras. Here again, slow-scan or time-sharing techniques can conserve channel bandwidth. For example, six cameras at six locations can be set to transmit sequentially, so that each location would be seen one-sixth of the time at the monitoring station.

Such services are not difficult technologically. Whether they are socially desirable and worth the cost in equipment and monitoring personnel must be determined by each city.

High-Speed Data Transmission

As Table 10 indicates, a large number of business data services have been proposed for two-way cable systems. Businesses today use the telephone network for most of them. Credit checking, for example, can be done with a telephone call—either by requesting verbal clearance, or by using a credit card reader to interrogate a remote computer. Stock market quotation systems use the telephone network to update stock prices displayed on cathode-ray-terminals at brokers' offices. Computer time-sharing services offer access by telephone lines to stored programs that can compute taxes, solve engineering problems, or retrieve information from a central

---

[7] Picturephone, (R) under development by AT&T for use over the switched telephone network, requires only a 1 MHz bandwidth instead of 6 MHz. It gives a lower resolution picture but is perfectly satisfactory for many uses. Cable applications such as document display among libraries, fingerprint exchange among police precinct houses, and X-ray or EKG transmission among hospitals may adopt similar bandwidth compression or "slow-scan" techniques.

[8] See Yin, op. cit..

file. Hotel, airline, theater-ticket, and car rental reservations likewise are available by calling a central computer databank.

Whether cable TV offers a superior method of supplying such services must be evaluated on a case-by-case basis. Generally, however, cable is unlikely to compete effectively with the existing telephone network where point-to-point communications are involved. Moreover, the telephone network will be better able to serve small business and the general public until cable's penetration nears that of telephone service (92 percent of U.S. residences in 1972).

Conversely, cable's cost advantage in business applications emerges when a high volume of information is to be exchanged at high speed among only a few users. For example, bank branches within a city might exchange data and update transactions on a dedicated cable channel. As the "checkless, cashless society" approaches, department stores and other retail outlets could tie in to record purchases as they occur. Large corporate computer users also might find cable an attractive transmission medium, especially if cable systems were interconnected to provide national point-to-point links. This may in fact come about in the 1980s through satellite interconnection of urban cable systems, or through interconnection agreements between cable systems and the new specialized common carriers that intend to compete with the established telephone companies and Western Union.

The use of cable systems for high-speed data transmission, however, depends on their reliability even more than cost. Today, business users complain about the telephone network's reliability and error rate. Cable networks *could* be designed for reliable, low-error-rate data transmission, but they presently are not.[9] Consequently, businesses may be slow to use the cable, preferring to wait until some proven record of reliability has been compiled. This inertia, while understandable, may defer the widespread use of cable for business services much longer than proponents have sometimes estimated.

[9] The principal problems are noisy components, lack of redundancy, and additive upstream noise. See I. Switzer, "The Cable System as a Computer Network," *TV Communications,* Vol. 9, No. 7, July 1972, pp. 78-85.

VII. TECHNICAL STANDARDS

The problems of establishing adequate technical standards and monitoring ca ble system performance for compliance with those standards have plagued almost every city granting a cable franchise. Cable technology generally is unfamiliar to city authorities, and the cable industry itself has not agreed on standards. Consequently, there are few if any guidelines to follow.

Past approaches have varied all the way from vaguely requiring "good quality pictures" with no detailed standards at all, to including a long list of technical specifications (some ambiguous, some unmeasurable, and some contradictory) in the franchise ordinance.

The confusion has been somewhat reduced by the 1972 FCC regulations, which now require certain technical standards in all new major-market systems. The standards are far from comprehensive, however, and pertain to performance characteristics only. Such factors as *reliability* are not covered, even though they determine in large part whether the system adequately fulfills its function.

FCC TECHNICAL STANDARDS

The FCC rules define four classes of cable TV channels. Class I comprises broadcast TV programs that the cable system receives and distributes to subscribers. Class II comprises cablecast programs that originate within the cable system (excluding those for which special decoding or unscrambling devices are required, as for pay-TV). Class III includes other one-way services such as pay-TV, facsimile transmission, or electronic mail delivery. Class IV applies to "reverse" communications, in which information flows from the subscriber to the headend or central distribution center.

Of these four classes, the FCC has established technical standards only for Class I channels. The reason for not setting standards for Class II, III, and IV channels is to encourage experimentation in developing new cable uses and services. The Class I technical standards are reproduced in the Appendix.[1] Broadly speaking, they are designed to insure that most subscribers receive a picture and sound of approximately the quality received off the air in an average reception area.

[1] The complete FCC rules are contained in Rivkin, op. cit.

It is important to realize that statistical rather than absolute concepts are involved. The standards do not guarantee that, if met, *all* subscribers will get excellent pictures, or even that picture quality will necessarily be the same at all points in the cable system. Localized sources of interference may degrade some subscribers' pictures even though the cable system meets FCC standards.

The minimum performance level also is set below that which many city viewers now receive off-the-air. As a result, most major market cable systems will have to exceed the FCC standards if they are to sell subscriptions in good reception areas. A local franchise can require more stringent technical standards if the local authority is prepared to enforce them itself.[2]

The complete lack of standards for Class II, III, and IV channels poses some problems for communities. Class II channels are already in use for local origination and public access. The public access channel is a particular source of technical difficulty, since many individuals or organizations using that channel prepare their own programs with inexpensive video cameras and tape recorders. Often such programs taped in the field fall below the quality level the cable operator or subscribers find acceptable. This factor alone can reduce subscriber interest in watching the public access channel.

Moreover, local program interconnection among cable systems may not be possible without technical standards. The FCC-mandated educational and local government channels, for example, fall into Class II. There is presently no assurance that a taped program prepared by a school district or government agency for distribution over one cable system would be compatible with the cablecasting equipment of a second cable system—even within the same city. Requiring such technical compatibility is now left to local authorities.

Some Class III and IV channels are also coming into use, but are not yet at the point where meaningful technical standards can be established. Pay-TV (Sec. VI) falls into Class III, while experiments with two-way communications (Sec. V) fall under Class IV.

WRITING AND ENFORCING LOCAL TECHNICAL STANDARDS

As an electrical system, cable TV must meet applicable state, county, and city building codes that specify construction standards, materials, grounding and shielding techniques, and so forth. The requirements of the National Electrical Safety Code, the Underwriters' Laboratories, and the American National Standards Institute may also be incorporated into the franchise ordinance by reference. Beyond these, communities may want to set local technical standards that exceed the FCC's requirements in three principal areas:

- Television signal quality
- Reliability and maintainability
- Performance monitoring and testing

---

[2] Letter from Sol Schildhause, Chief of the FCC Cable Television Bureau, to Western Communications Inc., August 11, 1972.

Signal Quality

"Signal quality," a term derived from broadcast TV usage, relates to the picture and sound delivered to the viewer's TV set. Five categories, "excellent," "fine," "passable," "marginal," and "inferior," were established in 1959 by the Television Allocation Study Organization (TASO), an industry group set up at the FCC's request. Although some quantitative characteristics (e.g., signal-to-noise ratio) are involved, these are largely subjective measures, based on the viewer's reaction to the picture he sees.

For cable TV systems, many franchise ordinances have required "high quality" or "good quality" without defining these terms specifically. This simply invites argument over whether the franchise requirement is being met.

The FCC standards for Class I channels define a minimum signal-to-noise ratio, signal level and frequency tolerances, and other parameters that determine signal quality. If these quantitative standards are achieved and maintained, the presumption is that quality will be at least acceptable to most viewers. This presumption should be examined carefully.

Almost every component of a cable system introduces some perturbation to the original signal, and many of the effects are cumulative. The major problems are:

- Signals leaving the broadcast transmitter may degrade by the time headend antennas pick them off the air.

- Headend signal processing equipment with its concentration of electronic components may introduce degradation.

- Each amplifier in the cable system introduces two undesirable effects, "noise" and "distortion." Electrical noise consists of random disturbances (e.g., "static") introduced by all components in the path of current flow. Distortion arises because amplifiers do not amplify all portions of a signal in the same proportion. Thus the amplifier output is a distorted version of the input. Both effects being cumulative, subscribers at the far end of a trunk or feeder cable will not receive as good a picture as those closer to the headend.

- All connections to the cable network, whether simple subscriber taps or local origination facilities, introduce discontinuities that increase noise or distortion. The cable itself can act as a giant antenna, picking up faraway radio signals or minute noise sources.

- Set-top converters and TV set tuners are particularly troublesome sources of interference. Problems are aggravated when a cable carries more than 12 channels, since the 12 VHF channel frequencies were originally chosen to minimize interference.

Consequently, a city that wants to require better signal quality than the FCC rules demand must pay close attention to all aspects of cable system design. Simply tightening the FCC's quantitative standards may not suffice.

Almost all of the above sources of noise and distortion become much more critical for two-way systems, because subscriber-to-headend messages enter the cable at hundreds of points. Poor connectors, faulty splices, or careless grounding may funnel intolerable noise into the system.

Many one-way cable systems can tolerate relatively sloppy design and construction features, but two-way systems cannot. This fact is of special concern for systems built to offer one-way services initially, but with the capacity to expand into two-way services at a later date. Unless the system is checked for two-way transmission at the outset, the expansion may require inordinate effort and cost.

Reliability and Maintainability

Reliability can be defined as the capability to maintain a consistent level of performance, once initially established. A system that works well when it is operative, but breaks down often, has good performance characteristics but poor reliability.

Maintainability is the ease with which a malfunction can be corrected, once it does occur. All complex electronic systems will have some failures, no matter how well designed, but good maintainability minimizes system "down time."

In the telephone, computer, and other technical industries, reliability and maintainability are defined quantitatively and written into specifications as a matter of routine. Reliability is usually quantified as a specified "mean-time-between-failure," and maintainability as "mean-time-to-repair." These are statistically averaged objectives against which actual performance is measured. They require precise definition of terms such as "failure," upon which there may be disagreement, but once defined they replace subjective terms with objective measurable ones.

Reliability and maintainability are becoming important for new major-market cable systems. Many projected nonbroadcast services will demand a level of reliability higher than the current standards of the cable industry. Business data communications, alarm and sensor monitoring services, and information retrieval cannot tolerate a high incidence of errors or equipment breakdown.

Even though these new services may not be offered for several years, if at all, franchising authorities must treat reliability as a current concern, because it is extremely difficult and expensive to "add" reliability to a system at a later date. Such basic construction and installation details as grounding, shielding, and the type of connectors used can intrinsically limit the reliability of the system. Thus, if high reliability is needed, it should be an inherent part of the original system design.

Two problems arise immediately. First, it is difficult to establish reliability and maintainability specifications for cable systems that are in a rapid stage of technological evolution. Second, many of the desirable specifications would be more costly than either the cable system operator or subscribers are currently prepared to pay for.

There are no quick solutions to these problems. Still, franchising authorities should include questions of reliability and maintainability in their consideratons. Otherwise, they may foreclose many of the broadcast services that cable hopes to offer urban communities in the years ahead.

Performance Monitoring and Testing

Once technical performance or reliability standards have been established, the tasks of monitoring and assuring compliance follow. Many past franchises have included no monitoring provisions at all. Others have specified some performance testing, but without enforcement or penalty provisions.

The FCC rules require system performance tests to demonstrate compliance with federal technical standards. Once each year, the operator must test his system at three widely separate points, including one at the far end of the network. The individual tests, procedures, and equipment used are the operator's responsibility, but they must receive FCC approval that they do, in fact, make all the measurements necessary to assure compliance.

The FCC requirements represent a large step forward in establishing a common baseline for major-market systems and guaranteeing that federal technical standards will be monitored annually.[3] A number of uncertainties remain, however. First, testing a complex system at three points may not be a completely adequate indication of performance, particularly if the same three points are always selected. Second, there is no requirement for FCC or city personnel to supervise or even witness the tests. Third, the system presumably is acceptable if it passes the tests *once* during the year, regardless of its record of malfunction or reliability. To use an extreme and unrealistic example, a system meeting the standards on the day of the test would satisfy the FCC even if it were inoperative the rest of the year. Finally, no penalties are specified except the implication that tests would be repeated if technical standards are not being met. Presumably, the FCC could withdraw its certification of the system and thus deny it the right to carry broadcast signals, but this seems likely only as a last resort.

The question of appropriate penalties for noncompliance is of great concern to cities. In most past franchises, the only penalty provision was the threat of franchise cancellation. This "ultimate sanction" not only is inappropriate for minor system deficiencies, but usually does not even serve as a credible threat. To cancel a franchise, a city must establish a sound legal case of default on franchise requirements. This will be very difficult if the requirements themselves are imprecise and will, in any event, involve considerable delay. Cancellation is a last resort, then; it is next to useless in more normal situations where standards are partially met, or where minor improvements are needed.

If a city intends to write and enforce its own technical standards, a better approach is to specify the tests to be performed, as well as a range of compliance and penalty provisions. Tests should be conducted on a statistical sampling basis, with a large enough sample to demonstrate performance throughout the system (50 to 100 subscribers should be enough for a large system). At the simplest level, the test might consist of tuning the subscriber's TV set or converter through all channel positions and judging picture quality according to the subjective TASO standards. Other, more quantitative tests can be devised for nonbroadcast services. The cable operator might perform the tests and certify them to the city. Alternatively, they could be conducted by outside consultants or by the city's own staff in much the same manner as building inspection now takes place. New York City recently established an Office of Telecommunication with a full-time director and staff to monitor cable system performance and handle subscriber complaints. Other cities are considering creating similar offices, whose expenses presumably would be provided out of franchise fees.

Table 12 illustrates this approach (in oversimplified form). A set of tests would be performed at the designated subscribers' TV sets, or at equivalent terminals connected to the system. Some percentage of substandard service would be allowed

---

[3] The cable industry asked for and received a one-year stay of these performance testing requirements in 1972. They remain in effect for 1973 and beyond.

Table 12

EXAMPLE OF FRANCHISE COMPLIANCE AND PENALTY PROVISIONS:
10% SUBSTANDARD SERVICE ALLOWABLE

% Terminals Not Meeting Specifications	Initial Test: Days to Implement Corrective Action	Retest: Fine for Noncompliance
0-10	Not required	---
10-20	60	$ 1,000
20-30	90	5,000
30-40	120	25,000
Above 40	(a)	(a)

[a]Consider franchise cancellation.

(arbitrarily shown as 10 percent in Table 12). Thus if 45 out of 50 terminals passed the tests, no corrective action or penalty would apply. If more than 10 percent of the terminals tested did not meet specifications, the system operator would be given time to take corrective action. Retests would then be made. If the level of service were still substandard, the city would assess a series of fines (or rate reductions) based on the overall performance level. It might also specify that the system again be retested every six months until performance improved.

A similar schedule could be devised for system reliability and maintainability, perhaps giving the operator a certain amount of time to correct specified malfunctions. With this approach, the penalty seems more appropriate to the offense, and the threat of franchise cancellation is reserved for those situations where the city really means to carry it out.

PROBLEM AREAS

Technical standards provide a formalized response to the question, "*How good should a cable TV system be?*" To arrive at any sensible answer, another question must first be posed: "*What is a cable TV system supposed to do?*"

Conventional cable systems supply entertainment to subscribers by redistributing broadcast TV (and sometimes radio) programs. This has been the only source of revenue in the past and promises to be the major (if not the only) source in the immediate future. Consequently, commercial cable operators design their systems today primarily for entertainment services. They naturally resist additional technical requirements that will increase system costs without offering compensating revenues. They often conclude it will be financially advantageous if they wait till later to install new equipment and tighten system specifications, even if it costs much more to retrofit than to build in these features at the start. Businessmen place a greater time value on money (i.e., discount rate) than do most government officials or policy analysts.

Cities and community organizations, on the other hand, may place priority on nonentertainment services. They may want technical standards written to insure that the cable system can provide additional education channels, data links among city hospitals, and interactive services to the home. Their interest has been fueled by years of glowing predictions of cable's vast potential for improving urban communications.[4]

The FCC regulations bring some comfort to both views, and also some confusion. Requiring that 50 percent of the total channel capacity be reserved for nonbroadcast use is an attempt to encourage major market cable TV systems to provide new, nonentertainment services. At the same time, establishing technical standards only for Class I channels is no help to cities that might want to lay at least the groundwork for implementation of new services in the future. Introduction of new services may, in fact, be delayed if the initial system design is based only on the FCC standards.

The heart of the problem is that a cable system good enough for TV entertainment may simply not be good enough for many nonentertainment services.

Communities have no easy solutions to turn to. Since most new services are, by definition, untested at this time, franchising authorities cannot specify a cable system design that will meet all future requirements. The more flexibility and the more stringent standards they require, the more the system will cost above one oriented primarily toward entertainment services.

Most cities have not faced the question of who will pay this additional cost. Since many proposed new applications will serve education, health care, public safety, and other public needs, a city should be prepared to share the cost of system upgrading with the cable system operator. If the cost burden is apportioned equitably, a city can then in good conscience expect and specify the levels of performance and reliability these new functions will require.

If, on the other hand, the added system cost cannot be justified initially, franchising authorities should at least be aware that many future services may be diminished in scope or foreclosed entirely. These considerations should be explored during planning and franchising, not after a franchise award.[5]

Even without considering the added cost of a higher quality system, the cost of monitoring a cable system for compliance with technical standards can easily exceed city revenue from franchise fees. A 20,000-subscriber system with subscriber rates of $72 per year will gross $1,440,000 a year. At the FCC-preferred rate of 3 percent, the city's annual franchise fee would amount to $43,200. This might support only one or two full-time technicians, with equipment, supervision, and overhead added. Consequently, cities must carefully add up the costs and benefits of an active local regulatory program, lest its cable system become an unforeseen revenue drain rather than a new source of income.

---

[4] For example, see the report of the Committee on Telecommunications, *Communications Technology for Urban Improvement*, National Academy of Engineering, Washington, D.C., 1971.

[5] For further discussion of the planning and franchising processes, see Baer, op. cit., Chapter 5, and Chapters 2 and 4 of this book.

Appendix

FCC TECHNICAL STANDARDS FOR CABLE SYSTEMS

Subpart K—Technical Standards

§ 76.601 Performance tests.

(a) The operator of each cable television system shall be responsible for insuring that each such system is designed, installed, and operated in a manner that fully complies with the provisions of this subpart. Each system operator shall be prepared to show, on request by an authorized representative of the Commission, that the system does, in fact, comply with the rules.

(b) The operator of each cable television system shall maintain at its local office a current listing of the cable television channels which that system delivers to its subscribers and the station or stations whose signals are delivered on each Class I cable television channel, and shall specify for each subscriber the minimum visual signal level it maintains on each Class I cable television channel under normal operating conditions.

(c) The operator of each cable television system shall conduct complete performance tests of that system at least once each calendar year (at intervals not to exceed 14 months) and shall maintain the resulting test data on file at the system's local office for at least five (5) years. It shall be made available for inspection by the Commission on request. The performance tests shall be directed at determining the extent to which the system complies with all the technical standards set forth in § 76:605. The tests shall be made on each Class I cable television channel specified pursuant to paragraph (b) of this section, and shall include measurements made at no less than three widely separated points in the system, at least one of which is representative of terminals most distant from the system input in terms of cable distance. The measurements may be taken at convenient monitoring points in the cable network: *Provided,* That data shall

be included to relate the measured performance to the system performance as would be viewed from a nearby subscriber terminal. A description of intruments and procedure and a statement of the qualifications of the person performing the tests shall be included.

(d) Successful completion of the performance tests required by paragrapll (c) of this section does not relieve the system of the obligation to comply with all pertinent technical standards at all subscriber terminals. Additional tests, repeat tests, or tests involving specified subscriber terminals may be required by the Commission in order to secure compliance with the technical standards.

(e) All of the provisions of this section shall become effective March 31, 1972.

§ 76.605 Technical standards.

(a) The following requirements apply to the performance of a cable television system as measured at any subscriber terminal with a matched termination, and to each of the Class I cable television channels in the system:

(1) The frequency boundaries of cable television channels delivered to subscriber terminals shall conform to those set forth in § 73.603(a) of this chapter: *Provided, however,* That on special application including an adequate showing of public interest, other channel arrangements may be approved.

(2) The frequency of the visual carrier shall be maintained 1.25 MHz±25 kHz above the lower boundary of the cable television channel, except that, in those systems that supply subscribers with a converter in order to facilitate delivery of cable television channels, the frequency of the visual carrier at the output of each such converter shall be maintained 1.25 MHz±250 kHz above the lower frequency boundary of the cable television channel.

(3) The frequency of the aural carrier shall be 4.5 MHz±1 kHz above the

65

frequency of the visual carrier.

(4) The visual signal level, across a terminating impedance which correctly matches the internal impedance of the cable system as viewed from the subscriber terminals, shall be not less than the following appropriate value:

Internal impedance:
75 ohms.
300 ohms.
Visual signal level:
1 millivolt.
2 millivolts.

(At other impedance values, the minimum visual signal level shall be $\sqrt{0.0133}$ Z millivolts, where Z is the appropriate impedance value.)

(5) The visual signal level on each channel shall not vary more than 12 decibels overall, and shall be maintained within

(i) 3 decibels of the visual signal level of any visual carrier within 6 MHz nominal frequency separation, and

(ii) 12 decibels of the visual signal level on any other channel, and

(iii) A maximum level such that signal degradation due to overload in the subscriber's receiver does not occur.

(6) The rms voltage of the aural signal shall be maintained between 13 and 17 decibels below the associated visual signal level.

(7) The peak-to-peak variation in visual signal level caused undesired low frequency disturbances (hum or repetitive transients) generated within the system, or by inadequate low frequency response, shall not exceed 5 percent of the visual signal level.

(8) The channel frequency response shall be within a range of ±2 decibels for all frequencies within −1 MHz and +4 MHz of the visual carrier frequency.

(9) The ratio of visual signal level to system noise, and of visual signal level to any undesired cochannel television signal operating on proper offset assignment, shall be not less than 36 decibels. This requirement is applicable to:

(i) Each signal which is delivered by a cable television system to subscribers within the predicted Grade B contour for that signal, or

(ii) Each signal which is first picked up within its predicted Grade B contour.

(10) The ratio of visual signal level to the rms amplitude of any coherent disturbances such as intermodulation products or discrete-frequency interfering signals not operating on proper offset assignments shall not be less than 46 decibels.

(11) The terminal isolation provided each subscriber shall be not less than 18 decibels, but in any event, shall be sufficient to prevent reflections caused by open-circuited or short-circuited subscriber terminals from producing visible picture impairments at any other subscriber terminal.

(12) Radiation from a cable television system shall be limited as follows:

Frequencies	Radiation limit (microvolts/ meter)	Distance (feet)
Up to and including 54 MHz....	15	100
Over 54 up to and including 216 MHz.	20	10
Over 216 MHz..................	15	100

(b) Cable television systems distributing signals by using multiple cable techniques or specialized receiving devices, and which, because of their basic design, cannot comply with one or more of the technical standards set forth in paragraph (a) of this section, may be permitted to operate provided that an adequate showing is made which establishes that the public interest is benefited. In such instances the Commission may prescribe special technical requirements to ensure that subscribers to such systems are provided with a good quality of service.

(c) Paragraph (a) (12) of this section shall become effective March 31, 1972. All other provisions of this section shall become effective in accordance with the following schedule:

	Effective date
Cable television systems in operation prior to March 31, 1972..................	Mar. 31, 1977
Cable television systems commencing operations on or after March 31, 1972......	Mar. 31, 1972

§ 76.609 Measurements.

(a) Measurements made to demonstrate conformity with the performance requirements set forth in §§ 76.701 and 76.605 shall be made under conditions which reflect system performance during normal operations, including the effect of any microwave relay operated in the Cable Television Relay (CAR) Service intervening between pickup antenna and the cable distribution network. Amplifiers shall be operated at normal gains, either by the insertion of appropriate signals or by manual adjustment. Special signals inserted in a cable television channel for measurement purposes should be operated at levels approximating those used for normal operation. Pilot tones, auxiliary or substitute signals, and nontelevision signals normally carried on the cable television system should be operated at normal levels to the extent possible. Some exemplary, but not mandatory, measurement procedures are set forth in this section.

(b) When it may be necessary to remove the television signal normally carried on a cable television channel in order to facilitate a performance measurement, it will be permissible to disconnect the antenna which serves the channel under measurement and to substitute therefor a matching resistance termination. Other antennas and inputs should remain connected and normal signal levels should be maintained on other channels.

(c) As may be necessary to ensure satisfactory service to a subscriber, the Commission may require additional tests to demonstrate system performance or may specify the use of different test procedures.

(d) The frequency response of a cable television channel may be determined by one of the following methods, as appropriate:

(1) By using a swept frequency or a manually variable signal generator at the sending end and a calibrated attenuator and frequency-selective voltmeter at the subscriber terminal; or

(2) By using a multiburst generator and modulator at the sending end and a demodulator and oscilloscope display at the subscriber terminal.

(e) System noise may be measured using a frequency-selective voltmeter (field strength meter) which has been suitably calibrated to indicate rms noise or average power level and which has a known bandwidth. With the system operating at normal level and with a properly matched resistive termination substituted for the antenna, noise power indications at the subscriber terminal are taken in successive increments of frequency equal to the bandwidth of the frequency-selective voltmeter, summing the power indications to obtain the total noise power present over a 4 MHz band centered within the cable television channel. If it is established that the noise level is constant within this bandwidth, a single measurement may be taken which is corrected by an appropriate factor representing the ratio of 4 MHz to the noise bandwidth of the frequency-selective voltmeter. If an amplifier is inserted between the frequency-selective voltmeter and the subscriber terminal in order to facilitate this measurement, it should have a bandwidth of at least 4 MHz and appropriate corrections must be made to account for its gain and noise figure. Alternatively, measurements made in accordance with the NCTA standard on noise measurement (NCTA Standard 005–0669) may be employed.

(f) The amplitude of discrete frequency interfering signals within a cable television channel may be determined with either a spectrum analyzer or with a frequency-selective voltmeter (field strength meter), which instruments have been calibrated for adequate accuracy. If calibration accuracy is in doubt, measurements may be referenced to a calibrated signal generator, or a calibrated variable attenuator, substituted at the point of measurement. If an amplifier is

used between the subscriber terminal and the measuring instrument, appropriate corrections must be made to account for its gain.

(g) The terminal isolation between any two terminals in the system may be measured by applying a signal of known amplitude to one and measuring the amplitude of that signal at the other terminal. The frequency of the signal should be close to the midfrequency of the channel being tested.

(h) Measurements to determine the field strength of radio frequency energy radiated by cable television systems shall be made in accordance with standard engineering procedures. Measurements made on frequencies above 25 MHz shall include the following:

(1) A field strength meter of adequate accuracy using a horizontal dipole antenna shall be employed.

(2) Field strength shall be expressed in terms of the rms value of synchronizing peak for each cable television channel for which radiation can be measured.

(3) The dipole antenna shall be placed 10 feet above the ground and positioned directly below the system components. Where such placement results in a separation of less than 10 feet between the center of the dipole antenna and the system components, the dipole shall be repositioned to provide a separation of 10 feet.

(4) The horizontal dipole antenna shall be rotated about a vertical axis and the maximum meter reading shall be used.

(5) Measurements shall be made where other conductors are 10 or more feet away from the measuring antenna.

§ 76.613 Interference from a cable television system.

In the event that the operation of a cable television system causes harmful interference to reception of authorized radio stations, the operation of the system shall immediately take whatever steps are necessary to remedy the interference.

§ 76.617 Responsibility for receiver-generated interference.

Interference generated by a radio or television receiver shall be the responsibility of the receiver operator in accordance with the provisions of Part 15, Subpart C, of this chapter: *Provided, however,* That the operator of a cable television system to which the receiver is connected shall be responsible for the suppression of receiver-generated interference that is distributed by the system when the interfering signals are introduced into the system at the receiver.

Chapter 2

The Process of Franchising

Leland L. Johnson and Michael Botein

I. INTRODUCTION

OBJECTIVES

The purpose of this report is to assist local governments and citizens groups in working through the various stages of the franchising process, in terms of both the mechanics and the policy issues involved in writing and awarding a well-conceived franchise. The focus throughout is on those jurisdictions that have yet to go through the franchising process.

Section II describes the various steps of the process, including the early delineation of franchise boundaries, the request for proposals (RFP), the steps for public hearings, the final award of a franchise, and the application for a certificate of compliance from the Federal Communications Commission. Examples of actual franchise provisions illustrate the process of selection and award.

Section III concentrates on the many major and minor provisions that a franchise should include, taking into account the FCC's recent *Cable Television Report and Order,* as well as other recent rulings, which encompass rules and guidelines that have a far-ranging effect on choices open to local jurisdictions. Particular sections that should be included in a franchise are discussed roughly in the order they would take in the final document. In each section, major policy issues are discussed in light of the new FCC rules; provisions that have been included in actual franchises are quoted to provide examples of both solutions to follow and pitfalls to avoid, as well as to guide the detailed drafting of the franchise document.

For illustrative examples the following franchises or ordinances are drawn upon: New York City (the Borough of Manhattan); Chicago; Los Angeles; San Mateo, California; Beverly Hills, California; Akron, Ohio; Pomeroy, Ohio; Columbus, Ohio; and Louisville, Kentucky. Useful guidelines are also to be found in the model franchise drafted by the Franchise Committee of the Miami Valley Council of Governments of the Dayton, Ohio area (MVCOG); the model ordinance of the National Institute of Municipal Law Officials (NIMLO); and the model code of the American Civil Liberties Union (ACLU).

An appendix contains a checklist of major considerations drawn from the preceding portions of the report. This provides a capsule view of important items that local authorities should keep in mind as they work through the details of the franchising process.

For many audiences, this report will be most useful when read in conjunction with several other documents, in addition to the summary Handbook:

- The FCC *Cable Television Report and Order* appearing in the Federal Register, Vol. 37, No. 30, Part II, February 12, 1972.
- The FCC *Memorandum Opinion and Order on Reconsideration of the Cable Television Report and Order*, appearing in the Federal Register, Vol. 37, No. 136, Part II, July 14, 1972.
- A companion report in this series by Steven R. Rivkin, *Cable Television: A Guide to Federal Regulations*, Crane, Russak & Co., New York, 1974, which also contains the full text of the *Cable Television Report and Order*.
- A second companion report, by Robert K. Yin, "Citizen Participation in Planning," Chapter 4 of this book.
- Two reports by the Cable Television Information Center: *How to Plan an Ordinance*(1972), which draws on existing franchises to provide additional illustrative examples of alternative ways to draft provisions; and *A Suggested Procedure*(1972), which outlines suggested steps in the local authorization of cable television.

A NOTE ON SOURCE MATERIALS

Although this study centers on local regulation of cable television, the FCC's rules affect many aspects of local regulation. As a result, an understanding of the local regulatory role requires at least some familiarity with the requirements—and often the ambiguities—of the federal rules. Both layman and lawyer, however, may find the rules' source as elusive as their substance.

The FCC promulgates a rule by means of a *Report and Order.* Though usually printed as one document, it is actually two. The *Order* adopts the set of rules, while the *Report* contains the FCC's interpretation of the rules, and its reasoning in adopting them.

A number of sources for any FCC document exist. The most important is the *Federal Register,* an official daily publication that prints all major federal documents. Under the federal Administrative Procedure Act an FCC rule usually cannot become law until it has been published there. Moreover, the *Federal Register* usually is the easiest source for FCC documents, since most libraries carry it and since copies of individual issues are usually obtainable by writing to the Government Printing Office. Because of these factors, this study gives the *Federal Register* as the source for FCC policy statements and regulations. Two other sources also exist: the official *FCC Reports* and the privately published *Pike & Fischer Radio Regulation.* Though both of these sources are accurate, they usually are harder to obtain than the *Federal Register.*

After a rule has been published in the *Federal Register,* it receives a section number in the *Code of Federal Regulations.* The *Code* is reissued every year and contains a permament and comprehensive collection of all regulations of federal agencies. Each regulation is classified according to a title and section number. All FCC regulations carry a title number of 47, since the FCC's regulatory power comes from title 47 of the United States Code. For example, since the FCC has assigned its rule on access to cable television a section number of 76.251, the rule can be found in 47 C.F.R. § 76.251 (1972). For convenience in referencing, this study gives not only the appropriate C.F.R. source, but also the *Federal Register* page on which the rule originally was printed, since revisions in FCC rules adopted in the last year will not have been codified.

More generally, FCC regulations and law review articles are referenced here in the standard legal format. All other literature is referenced in a conventional fashion consistent with that of other Rand reports.

II. DRAFTING AND AWARDING THE FRANCHISE

Franchising is a continuing and complex process, whose parameters are obviously less than exact. The process cannot be broken down into a neat "do it yourself" kit for franchising authorities and citizens groups to assemble one piece at a time. Nevertheless, it is possible to specify a generalized—albeit somewhat arbitrary—outline of alternative approaches and steps in the franchising process.

THE NEGOTIATION APPROACH VERSUS THE COMPETITIVE BID AND AWARD APPROACH

Two contrasting approaches exist in the franchising process with many variations between them. One is the "negotiation" approach, in which the city selects a prospective cable operator and negotiates the many terms and conditions that will go into the franchise. If the negotiations break down the city may, of course, elect to work with some other prospective grantee.

There are two major advantages to this approach: (a) it gives the franchising authority and the prospective grantee flexibility to agree tentatively on various provisions which can later be renegotiated depending on terms agreed elsewhere in the draft franchise, as an interactive trade-off process, so that an attractive overall package can finally be agreed upon, and (b) it expedites the franchising process insofar as relatively few participants are involved in the back-and-forth negotiations. Its major disadvantages are that (a) the early selection of a single prospective operator can give rise to complaints and even threats of lawsuits by other prospective candidates, (b) it does not permit the degree of community and citizen participation in the franchising process that such groups nowadays are demanding, and (c) however scrupulously the negotiations are conducted, the participants are vulnerable to suspicion by outsiders that under-the-table dealings are involved—all the more so since a number of illegal and questionable practices have come to light in the past.

Under the second approach, the "competitive bid and award" process, the franchising authority cooperates with community and citizen groups to decide in substantial detail the desirable elements of cable system design, ownership, and operation prior to dealing with prospective cable operators. The franchising authority issues its package of requirements and negotiable options on a fully competitive basis including requests for proposals, extensive public hearings, and formal evaluations prior to franchise award.

The advantages of this approach lie in avoiding real or alleged favoritism toward a particular prospective grantee, and in permitting community and citizen participation throughout the process. Its disadvantages lie in being (a) potentially more time-consuming, with the large number of steps and participants involved, (b) more costly in terms of additional staff and consulting resources required by the franchising authority and additional burdens placed on franchise candidates, and (c) less flexible, insofar as the franchising authority has greater difficulty in modifying its earlier decisions about terms and conditions (based on new information and other factors) once its request for proposals is issued.

Bearing in mind the many possible mixes of these two approaches, we shall discuss the detailed steps required in the competitive bid and award approach. In cases where franchising authorities prefer stronger elements of negotiation, some of the steps discussed below would be eliminated or compressed in time and scope. These steps include:

1. Adoption of procedures for drafting and awarding the franchise.
2. Hearings and tentative decisions regarding geographic area of each franchise and interconnection of franchisees within separate local jurisdictions.
3. Hearings on and adoption of a draft franchise document describing the general terms and conditions of the final franchise.
4. Preparation and dissemination of a request for proposals from franchise applicants, based upon the tentative franchise document.
5. Hearings on proposals received from applicants for the franchise.
6. Decision on award of franchise.
7. Application for an FCC certificate of compliance, including whatever special showings the franchisor and franchisee must make, either jointly or separately.
8. Continuing administration of the franchise.

STEP 1: ADOPTION OF PROCEDURES FOR DRAFTING AND AWARDING THE FRANCHISE

Initial adoption of a detailed procedural framework for awarding a franchise can allay citizens groups' fears of being "railroaded," and also reduce protracted and bitter procedural infighting later in the franchising process. Though the *Cable Television Report and Order* represents the FCC's most concerted and sophisticated effort so far to create standards for the local franchising process, a wide range of responsibilities exists at this nonfederal level. In terms of franchising, the FCC's only important procedural provision is the requirement that the franchising authority hold a "public proceeding affording due process" before choosing a franchisee or authorizing rate changes.[1] Under such a general requirement, the real responsi-

---

[1] 47 C.F.R. § 76.31(a) (1972), as printed in *Cable Television Report and Order,* 37 Fed. Reg. 3252, 3281 (1972); hereafter cited as *Cable Television Report and Order.*

bility for formulating a franchising procedure remains in the hands of state and local authorities.

Accordingly, there seem to be three main methods for adopting a procedural framework. First, a local franchising authority—or perhaps a consortium of local franchising authorities—might enter into a nonbinding "gentlemen's agreement" to follow certain procedures. Second, a local franchising authority—or again a consortium—might enact a procedure into local law by way of an ordinance; depending on local circumstances, an amendment to the city's charter or other basic enabling document might be preferable. Finally, a state statute might specify a uniform procedure for all municipalities. In reality, all three approaches may function in the same way; since no written instrument can hope to cover all possible contingencies, the ultimate determinant may be the good faith of the parties.

The procedural framework outlined above and detailed later provides for strong citizen input. Citizen participation is not only proper in drafting a well-conceived franchise, but also necessary as a matter of sheer political reality. As another report in this series notes in more depth, it is true that public input may lead to political infighting and thus delay the franchising process,[2] but today's citizen groups—particularly minority groups—demand that their voices be heard.

The procedural suggestions here would create three separate stages for citizen input: the initial planning, the drafting of a tentative franchise document, and the final choice of an applicant. These three separate steps actually should create a more efficient and speedy franchising process. If citizens are given a voice at each major juncture in policy decisions, it will not be necessary to backtrack at the final franchise award stage to consider basic policy matters.

In suggesting this procedural framework and outlining the subsequent activities, we do not set down appropriate time spans between steps. Aside from the fact that hard data on cities' actual experience are sparse, any attempt to generalize in this area would serve little purpose; the appropriate time frame for each municipality will vary with its size, demographic makeup, politics, and the like—as well as plain chance. Probably the only safe generalization is that a prolonged franchising process is ultimately preferable to an overly short one; hasty granting of "midnight franchises" has led some cities into ill-advised long-term commitments.

STEP 2: ASSESSMENT OF COMMUNITY NEEDS, OBJECTIVES, AND ALTERNATIVES

During step 2 the community would:

- Send for and review published information about cable;
- Review the experience of other communities;
- Conduct community surveys;
- Consider the size and characteristics of franchise districts, including possibilities of coordinated franchising with nearby jurisdictions;
- Investigate uses of the public access, education, and local government channels;

[2] R. K. Yin, "Citizen Participation in Planning," Chapter 4 of this book.

- Study ownership alternatives, system design concepts, economic implications, and other issues.

Some would call this the study phase. It can be as limited or extensive as the community wants and can afford.

Review of Published Information

A vast amount of published information about cable is now available—so much, in fact, that we seem more bent on papering the cities than wiring them. Guides to the literature, such as those listed in the Handbook references, can be helpful.

Review of the Experience of Other Communities

Direct conversations with local officials and others involved with franchising in other communities also will prove useful. Although individual word-of-mouth contact is the most tried and true approach, organizations such as the Cable Television Information Center, the National League of Cities, the International City Management Association, and state cable television associations can help locate knowledgeable people. These organizations, some universities, professional societies, and research institutions such as the Stanford Research Institute, the MITRE Corporation, and The Rand Corporation also hold conferences to enable municipal officials and others to exchange views on cable issues.

Staff members from the FCC Cable Television Bureau and the Cable Television Information Center may come to the community on request and talk with the franchising authority about specific local problems. This can be a valuable service; franchising authorities would be well advised to confine their requests to this purpose, rather than ask for introductory briefings on the general subject of cable television.

Conduct of Community Surveys

Surveys can provide helpful information for planning a cable system, such as data on television viewing habits, programming of special interest to particular groups, and the likely appeal of basic cable television service to the community. Surveys may also measure the potential appeal of new programs and services— whether more people would pay to see a local basketball game than to see the Metropolitan Opera, for example—though such preferences should not be construed as hard market data. Finally, surveys can indicate the present pattern of communications among residents, both within their neighborhoods and with people in other parts of the city. This information can provide some guidelines for the physical layout of cable trunk lines and for determining interconnection requirements among cable districts.

Surveys also give the ordinary citizen some voice in the planning process, albeit a passive one. If not properly designed and conducted, however, surveys will be of little value and may even yield incorrect results. Communities therefore should seek professional advice before beginning a survey project.

Franchise Boundaries and Interconnection

Two of the early major questions that confront the franchising authority are the geographic coverage of proposed franchises and the coordination of its franchising efforts with neighboring authorities. The two subjects are closely related, since the type of systems surrounding a city should influence the city's decisions about structuring its own franchises. For example, a small town's governing body would be justified in attaching less importance to an applicant's ability to originate programming if a large city's cable system were within interconnection distance.

First, a local franchising authority must decide whether to grant the whole franchise area to one operator or to carve it up into separate sectors with different operators. Appendix B discusses in some detail the pros and cons of single and multiple ownership of cable systems within a given city. Because of the complexity of the issues, it should be studied carefully before reaching conclusions.

In weighing the considerations of single versus multiple ownership, local authorities must keep in mind that much depends on the population of the area to be franchised. For a small city—e.g., 20,000 to 50,000 people—the investment requirements would be low for either a single operator or a set of operators; the economies of a single integrated operation would be large, and the problem of local control for such a small city relatively minor. The reverse may be true for cities with populations of several hundred thousand.

In any event, tentative decisions need to be made early to establish the franchise boundaries at least roughly (if the overall area is to be subdivided) and to give competing franchise applicants indications of the size of the market, the investment, and other requirements likely to be forthcoming. As additional information comes to light on the basis of the succeeding steps, these decisions may have to be reevaluated and modified.

Second, a local franchising authority must consider not merely the location of its own franchise areas, but also their relation to other existing or proposed systems in adjacent jurisdictions. One of the most disconcerting aspects of cable growth to date has been its fragmented geographic coverage. Franchises have been let separately by thousands of cities and counties scattered about the country, with each jurisdiction concentrating on franchising its own area. With little or no coordination among jurisdictions, cable systems have grown up side by side without interconnection to permit communication across jurisdictional boundaries.

No less serious, scattered about the interstices of municipal boundaries are unincorporated areas usually franchised by the county. In many cases, it is not economic for a cable operator to serve scattered clusters of houses or housing tracts adjacent to but not part of a municipality. It might be quite feasible to serve them with cable lines extended from the municipality, but this would require a cooperative arrangement between county and municipality permitting the city franchisee to expand beyond municipal boundaries.

So long as cable serves only to retransmit broadcast signals from a master antenna and conventional headend, as it has done in the past, the geographic fragmentation of the industry is not of great concern. For this service, cable construction is so straightforward that small communities can be independently served by separate operators. Indeed, fewer than 100 of the nearly 3000 cable systems operating today have more than 10,000 subscribers.

However, in order to realize the great promise of broadband communications, in contrast to the mere retransmission of broadcast signals, the nature of the industry will have to change. The use of computers, other sophisticated equipment, and high-quality (but high-cost) programming will be feasible only for subscriber bases numbering in the many thousands. New services—such as instruction in the home, shopping by cable, and information storage and retrieval—will be feasible only if a large geographic area can be simultaneously served from central points. Metropolitan-wide or regional networks of cable systems will be necessary to make such developments a reality. This, in turn, will require cooperation among franchising authorities—cities and counties—to permit a single system to serve more than one jurisdiction, or to permit separate systems to interconnect and provide large-area coverage while continuing to serve their own jurisdictions.

There are several mechanisms through which neighboring jurisdictions can seek agreements regarding requirements to be written into franchises for interconnection, compatible technical standards, and franchise boundaries. A voluntarily created "Council of Governments" representing the separate authorities in a metropolitan area may provide a suitable mechanism. A special cable television committee of the COG could be set up to investigate the benefits of alternative cooperative arrangements. While working out these arrangements, a voluntary freeze on awarding franchises among members of the COG may be desirable. In this process independent outside consultants may be brought in, and a brief study made of the economics and technology of cable in the particular environment. This study could draw heavily from the many previous studies in order to reduce costs and to ensure the use of tested methodologies.

Another possibility is to work through a local regional planning commission, if one exists, whose breadth of membership, interest, and political visibility may be sufficient to assist in working out satisfactory cooperative arrangements. Yet another alternative is for city governments, along with county representatives, to agree to set up a special ad hoc committee to study the situation and to make recommendations to the participating governments.

At the same time, all of these methods of cooperation among local governments may be mooted to the extent that states enter the regulatory arena—and an increasing number of states are expressing interest in doing so. Only a handful of states currently regulate cable; only one of these, Connecticut, has had substantial experience in allocating franchise areas on a region-wide basis. This limited experience suggests that state designation of franchise areas may well lead to a better distribution of cable services—although perhaps at the cost of considerable delay and confusion.[3]

Investigation of Uses of the Public Access, Education, and Local Government Channels

In major market cable systems, local government and community groups will be directly involved in using the three access channels mandated by the FCC. Much

---

[3] See M. R. Mitchell, *State Regulation of Cable Television,* The Rand Corporation, R-783-MF, October 1971. For an analysis of state regulation and a strong case in favor of state preemption of most local authority, see Barnett, *State, Federal and Local Regulation of Cable Television,* 47 *Notre Dame Law* 685 (1972).

of the planning effort should therefore focus on these uses. Two chapters of the summary Handbook deal with public access and public services on cable in more detail.

Community groups and interested citizens can be especially helpful in assessing the community's need for various services. A variety of devices—including surveys, neighborhood meetings, thorough studies, and committee reports—can contribute to the planning process. Videotaping individuals and groups who might utilize time on an access channel may be a particularly valuable technique. Local theater productions, neighborhood council meetings, and programming prepared by high school groups or senior citizens can be shown at public hearings or to the franchising authority. Such tapes can present an effective case for emphasizing public access in the franchise. They may also demonstrate that the community contains far more talent than it may have thought.

The franchising authority should consider delegating panels of community leaders, teachers and school administrators, and other citizens to come up with ideas and workable plans for use of the public access and education channels. It may also want to begin planning how the police, fire, social service, and other city agencies will use the local government channel. The plans developed for the use and administration of the access channels could then be discussed at the public hearings in step 3.

Study of Ownership Alternatives, System Design Concepts, Economic Implications, and Other Issues

A wide range of issues can usefully be brought out and analyzed by community groups and local government officials during the study phase. They can explore basic system design concepts, including the number and location of headends, trunk lines, and studios. An economic analysis of cable feasibility in the community might be made as described in Chapter 3 of the summary Handbook. Individual experts and outside study teams can be brought in as needed. Again, the objective should be to outline alternatives and set community priorities that can be reflected in franchise provisions.

Flexibility in Decisionmaking

Decisions about franchise boundaries and intergovernmental cooperation and the other issues discussed above may need modification as information accumulates during the succeeding steps. Because franchising is a continuing process, however, decisions on franchise boundaries and intergovernmental cooperation must anticipate their potential effects on the later performance of the systems. For example, a decision to include sparsely populated areas within the franchise area will increase total wiring costs and thus require higher installation and service rates. A decision not to cooperate with an adjoining jurisdiction ultimately may deprive a government's constituents of programming originated by another system.

Here, as elsewhere, franchising is not an orderly sequence of easily defined decision points, but an iterative process with feedback, forward and backward. It should operate within a framework of decisionmaking flexible enough to adapt to new perceptions of problems, opportunities, and needs along the way.

This iterative process cannot take place in a vacuum; it requires thoughtful and detailed citizen input. But in turn, citizen input cannot take place unless it has formal and clear channels. The procedural framework called for in Step 1 must enable citizens' voices to be heard, even at this early stage.

STEP 3: HEARINGS AND TENTATIVE DECISIONS REGARDING MAJOR ISSUES

Procedures for Giving Public Notice

Potentially interested citizens in the franchise area must be effectively notified. Adequate notice is particularly important at this early stage, since a city's mere decision to grant a franchise is likely to attract little attention from the news media. Traditional forms of "legal notices" are also inadequate, since they usually notify only lawyers—and a rather small class of lawyers, at that.

Improvements along the following lines should be considered. First, the franchising authority could be required to print a notice of a specified size and format —e.g., three columns wide by four column inches long—on the television page of all local newspapers. An even more effective approach would be to place spot advertisements on all local television stations—a procedure somewhat akin to the FCC's requirement that stations broadcast notice of a pending renewal application. This approach, however, might be unduly difficult in terms of both drafting and administration. The intangible values of various program times on different stations would complicate the creation of any written formula; and the cost of preparing an effective advertisement, even if the time were donated as a public service announcement, might be unduly high. If notice is intended to reach citizens interested in television, this type of approach would nonetheless be the most effective—except for those citizens whose tastes lead them to "tune out" contemporary television.

As a second alternative, the franchising authority might be required to print notices in a specified format and post them in all areas it controls—city offices, municipally owned bus lines, and the like.

Both these approaches would impose some cost on the franchising authority, like other functions such as voter registration and tax collection. Although the cost could probably be recouped from franchise applicants by an initial application fee, the franchising authority will need to consider both the benefits and the costs (however covered) in deciding how far to carry the process of notification.

Whatever form of notice is used, its timing is also highly important. First notice should be given sufficiently in advance of any hearings—e.g., at least the traditional thirty days—to enable citizens to prepare intelligent and useful comments. As the deadline approaches, the notice should be repeated, but perhaps less extensively to spare the franchising authority undue expense.

In addition, there should be special efforts in the community to notify organizations that have an interest in the shape of a franchise. The notice might be accompanied by a specific set of questions the organization might wish to consider, and a list of suggested readings. The services of the city attorney or other informed officers to help the organization prepare its position should also be offered.

Public Hearings

After notice has been given, the franchising authority should be required to hold hearings that encourage meaningful citizen participation. The procedural framework might provide for hearings in the evenings and on weekends in order to allow participation by working people. Moreover, the franchising authority should be required to hold hearings in all different neighborhoods of the franchise area. For example, each hearing might be held in a different state legislative district, chosen at random. It goes without saying that an insufficiently publicized hearing, or one held in too small a room, is hardly a "public proceeding." If effective citizen participation is to be encouraged, the procedural framework should specify that all hearings must be held at locations with seating capacity for a given number of persons and facilities for accommodating the electronic and print media.

The hearings should be designed to provide a useful focus on particular issues. When the FCC held hearings on its rules, for example, it established panels on such issues as public access and forms of ownership. The panels were so constructed that competing views on the particular issue were sure to be represented.

The problem here—as with any type of hearing—is to insure that all voices are heard, but not to the point of unproductive repetition and wearisome tedium. Again, several approaches are possible. First, a time limit (e.g., five minutes) could be imposed on all interested parties. The obvious problem here is possible unfairness to groups that represent large numbers of citizens. Second, all parties could receive a minimum amount of time, subject to increases based upon the number of signatures they could solicit on a petition; a sliding scale might even be used—e.g., an extra five minutes for every hundred signatures. Third, the "one man, one vote" constitutional standard might be transformed into a "one man, one voice" requirement. Under this procedure, all parties desiring more than the minimum time might be required to submit petitions with signatures at some reasonable time before the hearing (e.g., twenty-four hours); each party's time allocation then would depend on the proportion between his number of signatures and the total of all signatures. Rebuttal time also would be necessary to insure the ventilation of all views; it could be set, however, simply as a percentage of the initial presentation time to which a party was entitled.

Explanation of Decisions

The franchising authority should be required to publish a written explanation of its decision about franchise boundaries and interconnection with neighboring areas, and about other issues, within a reasonable time—perhaps thirty days—from the time that it is rendered. Though experience under other procedural frameworks has shown that the requirement to explain a decision does not guarantee better-reasoned decisionmaking, a written decision at least gives interested parties positions they can argue for or against. As with the notice requirement, defining adequate publication of the decision is difficult. Nevertheless, the franchising authority should be required to use any or all of the suggested forms of notice to inform the public that the decision exists, and a specified number of copies should be required to be made available at designated municipal offices during specified hours.

Other Possibilities for Citizen Input

Provision for citizen input through the formalistic process of notice and hearings does not begin to exhaust the possible forms of citizen participation. Legal draftsmanship has severe limits, however, and cannot provide for many significant and perhaps ad hoc forms of input. As another report in this series points out, a whole range of other techniques—such as community surveys, conferences, committees, and quasi-official groups—can aid greatly in identifying and articulating citizens' desires.[4]

STEP 4: HEARINGS ON AND ADOPTION OF DRAFT FRANCHISE

Having decided upon the franchise's geographic area in cooperation with neighboring jurisdictions and in light of other major issues, the franchising authority should move on to consider the substantive terms of the franchise. Its objective is to decide which provisions must be included in the final franchise or franchises and which provisions are to be left open for consideration by the applicants. Again, citizen input would be useful. By this point, however, at least some citizens in the community will probably be aware of the franchise proceedings and their importance; moreover, the issue may have become important enough to draw some media attention. Consequently, notice and hearings should be provided for, but need not be as extensive as those in the preceding step.

In the process of drafting a franchise, the local government may wish to consider franchising a noncommercial, joint commercial and noncommercial, or municipal operator. Since only a handful of noncommercial and municipal systems currently exist, it is difficult to evaluate the advantages and disadvantages of each approach. Nevertheless, some preliminary comments can be made. A noncommercial system might have difficulty in raising venture capital unless it has an obviously attractive franchise area with little risk of failure, or unless it has good prospects of support from foundations interested in promoting the growth of cable. A joint venture composed of a noncommercial and a commercial corporation might be able to obtain funds more easily. Structuring the joint venture to avoid conflict would be difficult, however, and the combination of such diverse groups could create complex and possibly unfavorable tax consequences. A municipal system with a good credit rating would encounter no difficulty in raising capital, if the city had not reached its constitutional debt limit, or were not otherwise prevented from financing its system with a bond issue. In terms of public service, however, none of these three alternatives has a clear advantage over a purely commercial operation. Noncommercial programmers are as human, and can be as biased, as commercial ones; indeed, the prospect of turning over a broadband communications system to members of a local government creates possibilities of political abuse. Consideration of these points is an important area for citizen input.

On the basis of the preceding, the franchising authority should draft the major provisions of the franchise document as they will appear in final form. The basic

[4] Yin, op. cit.

analysis and drafting at this stage should consider the substantive issues discussed in Section III of this study. Some items must be left open since they will be determined by competition among prospective grantees. At the same time, the draft franchise should indicate at least the upper and lower bounds that the franchising authority would be willing to accept—e.g., a maximum monthly service rate of $6 or a requirement of particular program origination facilities—within which citizens can weigh their preferences. For example, strong preferences might be expressed for sacrificing a certain amount of program origination as a tradeoff for lower monthly subscriber rates. In any event, many of the legal considerations, definitions of terms, and other provisions should be included approximately as they will appear in the final document. As in Step 2, the franchising authority should be required to publish a written explanation of its decision.

STEP 5: PREPARATION AND DISSEMINATION OF REQUEST FOR PROPOSALS

Having drafted a tentative franchise document, the franchising authority should transform it into a formal request for proposals (RFP). This will include not only the proposed terms of the final franchise, but also a request for detailed information from the applicant about its financial condition, ownership interests, and proposals concerning provisions left open in the tentative franchise document.

Wide dissemination of the RFP is advisable to prevent limitations on a franchising authority's field of choice through either inadvertent omission or deliberate exclusion. The procedural framework should not simply designate a list of operators who should be invited to apply for the franchise; to achieve the widest possible dissemination, the franchising authority might be obliged to mail a copy of the RFP to any potential applicant designated by any citizen. As a less onerous alternative, the franchising authority might be required to furnish a copy of the RFP to any citizen and to accept for its consideration any properly completed proposal it receives from such a source.

Noncommercial groups, as discussed in Step 3, should be included in the distribution along with commercial organizations. If the city is interested in muncipal ownership, it could apply for the franchise as one of a number of applicants.[5]

The following example of some of the threshold requirements in a request for proposals is drawn from the Louisville ordinance:

A. No CATV Franchise or renewal thereof shall be issued except on a written application and upon a form approved by the Board of Aldermen as recommended by the CATV Special Committee of the Board. Such applications shall be accompanied by a nonrefundable cash fee of One Thousand Dollars ($1,000.00) which shall be paid to the Director of Finance.

B. The form shall set forth such facts in detail as the Board of Aldermen may deem appropriate including:

    1.   If the applicant is an individual, partnership, or unincorporated association, its statement shall contain the names and addresses of all persons (including corporations) having a proprie-

---

[5] The situation of the city competing with private parties obviously poses problems unless the franchising authority is in a position to deal at arm's length with *all* applicants. One of the most serious drawbacks to municipal ownership is the danger that it prematurely forecloses consideration of competing applications, once the city decides to own and operate the cable system.

tary or equitable interest in and to the prospective franchisee's business operation, and in and to the prospective franchise if awarded to the proposer. The term "equitable interest" shall include all assignment for value, as well as all contingent assignments of any right or privilege under the prospective franchise, and shall also include any benefit, payment, or emolument whatsoever resulting from the grant of a franchise under this ordinance.

2. If the applicant is a non-public corporation, the statement shall furnish, additionally, the names and addresses of the officers, directors, and shareholders of the said corporation, together with the number of shares held by each shareholder.

3. If the applicant is a publicly held corporation, as defined by the rules and regulations of the Securities and Exchange Commission, the statement shall contain the states in which incorporated and/or qualified to do business, the names and addresses of the officers and directors of the corporation, the names and addresses and number of shares owned of all stockholders both nominal and beneficial, owning 3% or more of the outstanding stock of the applicant and the names and addresses of each shareholder who is a resident of Jefferson County, Kentucky, or Clark or Floyd Counties, Indiana, together with the number of shares owned by each;

4. A full disclosure of the ownership of the facilities to be used in rendering the service;

5. The source of funds for operation of the system respecting the installation and maintenance of all CATV facilities; and shall demonstrate the financial ability to provide and extend service to proposed subscribers at a reasonable cost;

6. A detailed schedule of the rates to be charged for the services offered, the facilities to be employed and the general routes of the cables used in redistributing signals, the service area or areas, the commencement and completion dates of construction of the CATV System and the proposed dates the service will be available to the area or areas named.

7. A detailed schedule of rates to be applied to residential, apartment, commercial and other users of service.

C. The Board may request such other information as it may deem appropriate.

D. All applications shall be open to public inspection, shall be kept on file a reasonable length of time at the discretion of the Board, and any intentional misrepresentation in an application shall be grounds for its rejection or for termination of the franchise.

E. All applications shall be considered firm offers to the City, shall be signed and verified by the applicants whose relationship to the applicant shall be set forth and shall bind the applicant to the provisions thereof.

These terms of the Louisville ordinance suggest yet other provisions a franchising authority should consider. First, the flat cash application fee required in Subsection A may short-change the franchising authority; actual franchising expenses may far exceed preliminary guesses. A better approach might be to require applicants to deposit a larger amount of money in an escrow account. After the final franchise award has been made, the franchising authority could compute its actual costs and charge each applicant its pro rata share.

Alternatively, the bulk of the cost burden might be imposed on the applicant who is awarded the franchise, with full or partial refunds to the other candidates. When there is an elaborate selection mechanism, a high per capita entrance charge may discourage good applicants. True, the operator who builds the system would be able to pay a high entrance charge. But many organizations may be unwilling to risk, say a $10,000 to $15,000 charge when there are many competitors. By placing much of the cost burden on the applicant selected, the franchising authority may encourage a wider range of choice.

Second, Subsection B(2)'s required disclosure of a corporation's "officers, directors, and shareholders" is wise, since it allows both the franchising authority and the public to inquire into an applicant's background. This logic might be extended a bit further, by requiring the applicant to list any managerial or ownership interests that its "officers, directors, and shareholders" have in any other business. Similarly, Subsection B(3) perhaps should require disclosure of any persons who own less than three percent—perhaps even one percent—in a publicly held corporation;

even one or two percent of a large corporation can give its owner significant control. By contrast, the FCC requires disclosure of one percent interests in many situations. In addition, Subsection B(4)'s disclosure provisions concerning the "ownership of the facilities to be used" might be made more severe; leasing or buying land from a local official is one of the classic methods of exercising undue influence. Thus enumeration of each piece of property involved, as well as of its owner and his interests in other businesses might be more appropriate.

Third, Subsection B(6)'s requirement for information on the system's proposed rates, services, routes, etc., might be more specific. The franchising authority might request information concerning rates for leased channels as well as for subscribers. A description of routes should designate major streets and other ways by name or plot location. And perhaps most important, the information on extending service throughout the franchise area should be required to be in specific terms—e.g., state legislative districts or other existing boundaries within the city.

Finally, Subsection D's requirement of "public inspection" for a "reasonable length of time" raises several issues. First, "public inspection" is obviously a term susceptible to differing interpretations; it should be defined more precisely. For example, the franchising authority might be required to post copies of all proposals at designated spots in municipal buildings scattered throughout enumerated areas of the city. Second, "reasonable length of time" could be made more specific simply by substituting for it a posting requirement from the time the proposal is received to the time the final franchise award is made. In addition, the franchising authority might be required to make reproductions of all proposals on request at a specific cost within a specified time.

STEP 6: HEARINGS ON PROPOSALS

An illustration of problems in delineating the hearing process is drawn here from the Chicago ordinance:

2.4 *Hearings.* The Commission shall hold public hearings on all bids received. There shall be public notice one month prior to these hearings.

2.5 *Award.* All franchises shall be awarded within three months of the last day of public hearings. However, the Commission may, if all bids are unacceptable, call for new bids under the provisions of this Article.

2.6 *Publication.* Any franchise that is granted shall be published.

2.7 *Area Hearings.* All public hearings relating to a specific franchise area, including those held to consider the initial franchise grant, shall be held within the geographic limits of the designated area. At least half of the hearings in each franchise area shall be held at night or on Saturdays.

Though other sections of the ordinance define "public notice" and "publication" in sufficiently concrete ways, the ordinance does not specify how the "public hearings" will be held—thus creating the possibility that some voices will be heard and others not. The franchising authority might better ensure effective public notice and meaningful hearings by following procedures akin to those discussed in Step 2. In fact, the necessity for notice may be less pressing at this point than at the earlier stages, since the actual award of the franchise should attract considerable attention from the news media. The hearings should be conducted along the same lines as the previous hearings, with time allocated for presentations and rebuttals by franchise applicants.

STEP 7: DECISION ON AWARD OF FRANCHISE

This is one of the most difficult portions of the franchising process—particularly the selection of the applicant who proposes the package of elements, including a reasonable rate structure, that on the whole seems most attractive in light of the community's previously expressed desires.

In judging the reasonableness of proposed rates for services of the sort offered today (including charges for adding outlets, converters, and the like), the franchising authority will need to rely heavily on a close analysis of financial projections made by competing franchise applicants. These financial statements typically include annual profit and loss projections, a pro forma balance sheet, and other financial data, based on the estimated growth of the cable system over some period of time —typically ten years. Many examples of cable system cost inputs, financial statement format, and cost projections can be obtained from past cable studies.[6] The franchising authority should be careful, however, not to put itself in the position of simply balancing one applicant's proposals off against another's. Instead, the authority should have developed its own financial and service analysis before it even receives proposals, either using its "in-house" staff or, as often will be the case in smaller cities, through contracts with independent consultants.[7] The basic task, then, should be to measure each applicant's proposal against the authority's own analysis, in light of expressed public desires.

The following are some of the important questions the authority needs to ask:

- On what basis does the applicant estimate the annual growth and ultimate penetration of cable in this particular market?
- Do these estimates appear overly optimistic or pessimistic, considering the level of over-the-air broadcasting the cable operator will have to compete with in signing up viewers?
- Are revenues based on existing conventional services or do they include estimates of new services yet to be perfected? If the latter, does the applicant have a reasonable basis for estimating these revenues or are they pulled essentially out of the blue sky?
- Do the items of capital expenditures appear to be in the same ballpark as those estimated elsewhere? For example, is the cost per mile of cable plant low or high in comparison with experience in similar markets? If there is a substantial difference, is it justified in terms of the special conditions of the particular city or of the nature of the system being offered?

[6] See, for example, L. L. Johnson, W. S. Baer, R. Bretz, D. Camph, N. E. Feldman, R. E. Park, and R. K. Yin, *Cable Communications in the Dayton Miami Valley,* The Rand Corporation, R-943-KF/FF, January 1972, Paper 2; and Gary Weinberg, *Cost Analysis of CATV Components,* Resource Management Corporation, Report UR-170, June 1972.

[7] See Chapter 3 on cable system economics in W. S. Baer, *Cable Television: A Handbook for Decisionmaking,* Crane, Russak & Co., New York, 1974.

- How do the cost estimates of local program origination facilities (studios, cameras, lighting, and so forth) compare with estimates elsewhere?
- With respect to operating expenses, do payroll figures, pole rentals, property taxes (which, incidentally, are one of the largest single annual expense items), as well as selling and advertising expenses seem reasonable on the basis of experience elsewhere? Again, if not, do special circumstances of this franchise area explain the difference?
- Do the proposed debt-equity ratio and the rate of interest on debt seem appropriate, given the nature of the current capital market and the nature of the franchise applicant? Is the payback of debt and the flow of prospective dividends to stockholders estimated on grounds that would be regarded as financially prudent? If so, does this give a multiple system operator a preference that outweighs the desirability, if any, of local ownership and control?
- Is the estimated value of the system at the end of the period (a particularly important consideration in evaluating the viability of the system) reasonable in light of the market value of similar systems at the end of the same number of years? In the past, systems have typically been valued at from 7 to 10 times annual operating income. How does the estimate in this case compare with this range?

Of course, the applicant's financial qualifications are also important. General questions here include:

- Does the applicant have sufficient capital to build the system in accordance with his design and construction plan?
- If sufficient total funding is not available but some equity financing is in hand, is the borrowing requirement realistic in terms of the debt-equity ratio required for the plan?

These and other questions might best be appraised by a financial analyst engaged by the franchisor to undertake an across-the-board evaluation of the financial qualifications of all applicants. A major objective is to select the cable operator who will actually build the system, rather than one who merely hopes to obtain the franchise and sell it at a profit to some other, more responsible group. The franchisor must beware of a phenomenon that has occurred in the past, wherein groups of "leading citizens" join together to obtain a franchise but have no sound financial plan or real interest in exercising it—with the consequence that they sell the franchise (sometimes at a handsome profit) to another entity, which then proceeds to build the system.

Finally, in cases where the applicant has systems operating in other localities, the franchisor should check with authorities there to evaluate the applicant's track record. Here, a number of questions arise:

- How well did the applicant build and operate in accordance with the franchise?
- Did he have a solid financial base or was he obviously overextended?
- What was the level and nature of customer complaints? How well were they handled?

- Did the applicant pursue nondiscriminatory hiring and training practices?
- In general, was the franchisor satisfied with the performance of the operator? If not, what were the specific problems?
- How competent is the other franchising authority?

Answers to such questions will afford a better basis for comparing proposals. One applicant may offer a relatively low monthly rate—say $4—that is economic only if his visions of high advertising revenues come true. Another may propose a higher rate—say $6—but offer relatively elaborate local program origination facilities and a high quality signal. Yet another may offer a typical rate of $5—but with relatively high initial subscriber connection charges.

From these comparisons the franchise authority will have to make hard choices. Unfortunately, the complex tradeoffs involved preclude any neat formulas that supply ready answers. True, some proposals can be dismissed easily on grounds that the applicants' estimates of cable penetration are overly optimistic, or that their financial qualifications are weak. But this still will leave a "hard core" of proposals among which the final choice will be difficult. For example, is it better to have a $6 monthly subscription rate with relatively elaborate local origination facilities or a $5 rate with minimal ones? Is it better to have a "regular" $5.50 rate for every subscriber or to have a $6 rate that permits a special preferential $4 charge for the aged or poor?

Obviously, clear answers do not abound. Nevertheless, a few procedural guidelines at least can make the decisionmaking process more visible and perhaps more rational.

First, the franchising authority should designate and make highly visible the areas in which it will recognize competing bids. If it has decided to require local program origination facilities at several specified geographic locations but is flexible as to subscriber service rates, it should make clear that only the subscriber rates are subject to bidding. This approach not only will simplify the bidding for applicants, but also will make it easier for the franchising authority and the public to evaluate bids.

Second, the franchising authority should assign a definite weight and priority to each factor on which bids are taken. For example, it might give a weight of 10 percent to channel capacity, and 40 percent to subscriber service rates. It should then create priorities within each criterion—e.g., a $4 bid for the monthly subscriber service rate might be worth ten points, a $6 bid worth one. Though decisionmaking cannot be based solely on a mathematical formula, the franchising authority might find useful a chart, like that in Table 1, to help quantify and evaluate each applicant's proposal.

Table 1

ILLUSTRATIVE EVALUATION FORMAT

Applicant	Criterion	% of Decision	Score	Subtotal
A	Channel capacity	10	8	0.8
B	Channel capacity	10	6	0.6
A	Subscription rates	40	7	2.8
B	Subscription rates	40	5	2.0

Third, the procedural framework should require the franchising authority to write a reasoned opinion explaining its decision on the franchise award. As noted before in relation to previous steps, this requirement is difficult to enforce and easy to evade; nevertheless, the combination of citizen input, identification of bidding criteria, and evaluation of bids should all work toward more rational, or at least more candid decisionmaking.

These steps taken together are in marked contrast to past experience in the franchising process. Many cities have entered into long-term franchises without soliciting competitive bids, and many franchise proceedings have been conducted with little public notice that local governing bodies were considering cable franchises.[8]

STEP 8: FCC CERTIFICATE OF COMPLIANCE

The FCC's requirements for applications for certificates of compliance—without which a cable system may operate, but not carry any broadcast signals—are contained in 47 C.F.R. § 76.11. Though the rules appear complex at first glance, they actually require a straightforward listing of fairly general information. Nevertheless, several aspects of the rules are particularly important in the franchising process. Though the substantive as well as procedural requirements of the rules will be discussed in more detail in relation to the specific issues raised in Sec. III, some introductory observations are appropriate.

First, the rules specifically require the cable system to give formal notice to the franchising authority that it is seeking a certificate of compliance. As a result, the appropriate local official—usually the city attorney—should expect to receive a copy of the application for certification and should be prepared to take any action the local franchising authority deems necessary in supporting or opposing the application. This action must be taken without delay, however, since the rules give only 30 days from the FCC's public notice of the application in which to file objections. The application for certification also gives local citizens the chance to object to the conduct of the franchising process. If a citizen group feels that the requirement of a "public proceeding affording due process" has not been met, it may complain to the FCC at this point. Once again, however, the 30-day time limitation applies. Moreover, since the rules do not require the cable system to serve notice on citizen groups, they must move even faster than local franchising authorities.

Sometimes the local franchising authority must join with the cable system in applying for a certificate. As will be discussed in more detail in Section III, the FCC will permit a franchising authority to impose certain franchise terms only if the FCC approves them after a joint "special showing" by both the franchisee and the franchisor. As a result, the franchising authority must be prepared to engage in often complex administrative proceedings in order to secure some franchise terms. Citizen groups, of course, may object to special showings as well as to ordinary applications for certification.

---

[8] A number of specific examples are cited by Leone and Powell, *CATV Franchising in New Jersey,* 2 *Yale J. L. & Soc. Ac.* 254 (1972).

The 30-day time limitation is very important for citizen action. Although a general provision of 47 C.F.R. § 76.7 allows interested persons to petition the FCC for relief at any time, citizen groups cannot depend on this provision for making a tardy appeal. The FCC has shown that it will entertain petitions after the 30-day period only under the most extenuating circumstances. It will not be enough for a citizen group to say that it was not aware of the earlier proceeding, or that more time was required to marshal its forces and to make a presentation. Citizen groups can expect to use this route successfully only under the most compelling circumstances.

STEP 9: MONITORING SYSTEM CONSTRUCTION AND CERTIFYING PERFORMANCE

After FCC certification, the franchising authority must ensure that the cable system is constructed in compliance with the terms of the franchise. It will want to monitor the pace of construction and the hiring practices of the contractors, as well as the contractors' quality control procedures. In many cases this can be done by city inspectors, although engineering specialists will be needed for systems that are to deliver more than television services. As an example, expert monitoring will be necessary if the city expects to use the cable system for two-way communications among municipal agencies at an early date.

However simple or complex the system, its performance should be checked at several stages of construction. If performance tests are deferred until the system is ready to be "turned on" for subscribers, technical problems or variances from franchise terms may be difficult to correct. Instead, the franchising authority should make certain when the headend is completed that television signals delivered to the cable are of adequate quality. It might then check performance at the ends of the first feeder cables when they are installed. The franchise presumably will also include provisions for more formal "certification of performance" at the completion of each construction phase. The city may be able to certify performance on its own if it has the necessary equipment and technical staff. Most communities will hire a technical consultant for this task, however.

STEP 10: CONTINUING ADMINISTRATION OF THE FRANCHISE

Though the franchising and certificating processes are complex and difficult, their completion does not signal the end of local responsibilities. During the post-franchise period the cable operator, the franchising authority, and the community must live together on a day-to-day basis. The cable operator will have disputes with both citizens and the franchising authority; the franchising authority will have its own complaints against the cable system and will receive complaints from citizens; some citizens inevitably will be outraged at both the cable system and franchising authority. To compound matters, these conflicts may end up being resolved in any one of a number of forums. The courts will hear complaints from citizens and appeals from the franchising authority's decisions. The franchising authority will be called upon to resolve disputes between the community and the cable system. The FCC will receive objections to certification or petitions for special relief. Any major public endeavor generates conflict, but the stakes involved in the cable business virtually guarantee that it will generate more than its share. Section III will explore specific means of resolving particular types of conflicts.

III. TERMS AND CONDITIONS OF THE FRANCHISE

In writing the ordinance, drafting the tentative franchise document, and delineating terms of the final franchise, a number of provisions should be worded with particular care, to avoid the danger of subsequent disagreements, delays, and litigation. This section is organized around a number of topics, roughly in the order they would take in a completed franchise.

PREFATORY PROVISIONS

A franchise typically contains a set of "whereas and therefore" paragraphs setting down the process by which the franchise was granted and stating that the franchising authority and the grantee agree to all the conditions. An example is that of the New York City—i.e., The Borough of Manhattan—franchises held by TelePrompTer Manhattan CATV Corporation and Sterling-Manhattan CATV Company.

WHEREAS, by resolution adopted June 18, 1970 (Cal. No. 414), the Board of Estimate entered on its minutes an authorizing resolution and accompanying proposed contract and did fix July 23, 1970 as the date for a public hearing on said resolution and accompanying proposed contract; and

WHEREAS, on said date said Board duly held and closed such public hearing; now, therefore, be it

RESOLVED, that the Board of Estimate of The City of New York, hereby grants to Teleprompter Corporation and its subsidiary Teleprompter Manhattan CATV Corporation the franchise and right to install, operate and maintain a broadband communications facility, sometimes called a Community Antenna Television System, within a certain area in the Borough of Manhattan, upon and subject to all the terms and conditions contained in the accompanying contract, and that this resolution shall be duly certified and presented to the Mayor for his approval, and upon such approval, the Mayor of The City of New York be and he hereby is authorized to execute and deliver the accompanying contract in the name and on behalf of The City of New York, and that this resolution shall be null and void if Teleprompter Corporation and Teleprompter Manhattan CATV Corporation shall fail on their behalf to properly execute said contract in duplicate and deliver the same to this Board within Forty-five (45) days after the approval of this resolution by the Mayor or within such further time as the Board may grant by resolution adopted on a date prior to the expiration of said Forty-five (45) days.

This is a contract, executed in duplicate this 18th day of August, 1970 between The City of New York (the "City") by the Mayor of the City (the "Mayor"), acting in accordance with the authority of the Board of Estimate of the City (the "Board"), party of the first part, and Teleprompter Corporation and its subsidiary, Teleprompter Manhattan CATV Corporation, both organized and existing under the Laws of the State of New York (both hereinafter referred to as the "Company"), parties of the second part,

WITNESSETH:

WHEREAS, Teleprompter Corporation by petition dated October 19, 1964, applied to the Board for a franchise to install, operate and maintain a Community Antenna Television System; and

WHEREAS, said Board adopted a resolution on November 19, 1964 (Cal. No. 72), fixing the date for a public hearing on said petition as December 3, 1964, said petition and notice of public hearing thereon were duly published, and said hearing was held and continued to January 14, 1965 and closed on that date; and

WHEREAS, said Board adopted a resolution on December 2, 1965 (Cal. No. 128) authorizing Teleprompter Corporation to install, maintain and operate a Community Antenna Television System within a certain area comprising roughly the northern half of the Borough of Manhattan for an interim period terminating on December 31, 1967 (the "consent") and

WHEREAS, by a resolution adopted by said Board on June 10, 1966 (Cal. No. 44) consent was granted to the assignment of the consent to Teleprompter Manhattan CATV Corporation, a subsidiary of Teleprompter Corporation, and

WHEREAS, by modifying resolutions of November 22, 1967 (Cal. No. 130-A), December 19, 1968 (Cal. No. 41), December 18, 1969 (Cal. No. 243-A) and March 12, 1970 (Cal. No. 169-B) the consent was extended for periods expiring June 30, 1970 and the Company was granted permission to originate certain types of programs; and

WHEREAS, said Board has made inquiry as to the money value of the proposed franchise contract and the adequacy of the compensation proposed to be paid therefor; and

WHEREAS, said Board did embody the results of such inquiry in this contract and has caused this contract to be spread upon the minutes of the Board on June 18, 1970, together with the proposed resolution for the grant thereof and did fix the 23rd day of July 1970, for a public hearing thereon, at which citizens should be entitled to appear and be heard; and

WHEREAS, prior to said hearing, notice thereof and the proposed contract and proposed resolution authorizing this contract were published in full for at least fifteen (15) days (except Sundays and Legal Holidays) immediately prior thereto in the City Record and notice of such hearing, together with the place where copies of the proposed contract and resolution of consent thereto might be obtained by all those interested therein, was published at least twice, at the expense of the Company, in the two newspapers designated by the Mayor and said hearing was duly held and closed on said date;

Now, THEREFORE, the parties hereto do hereby mutually covenant and agree as follows:

Although such language may seem ceremonial, it does serve the important purposes of enumerating the procedural steps preceding the franchise grant and signifying that the franchise is in accordance with existing ordinances and rules. This helps to insure that the franchise stands up before the courts and the FCC, and to guide individuals and groups who are interested in examining the history of a particular franchise process.

DEFINITION OF TERMS

A list of clearly defined terms is important to avoid subsequent misunderstanding and disagreement. Deciding which terms to define will depend on the specific circumstances of the locality. A good example of definitions suited to local circumstances is contained in the ordinance of Louisville, Kentucky.

For purposes of this ordinance the following terms, phrases, words, abbreviations and their derivations shall have the same meaning given herein. When not inconsistent with the context, words used in the present tense include the future; words in the plural number include singular number; and words in the singular number include the plural. The word shall is always mandatory and not merely directory.

A.   *City* — shall mean the City of Louisville, Kentucky, a municipal corporation in the Commonwealth of Kentucky.

B. *Mayor* — shall mean the existing or succeeding chief administrative officer of the City, or his designate.

C. *Board* — shall mean the present governing body of the City or any successor to the legislative powers of the present Board of Aldermen.

D. *Director of Law* — shall mean the chief legal officer of the City of Louisville presently known as the Director of Law.

E. *Director of Finance* — shall mean the chief financial officer of the City of Louisville presently known as the Director of Finance.

F. *Director of Works* — shall mean the Director of Public Works of the City of Louisville.

G. *CATV* — shall mean Community Antenna Television.

H. *Community Antenna Television System (hereinafter called CATV System)* — shall mean any facility in which (1) in whole or in part receives directly or indirectly over the air and amplifies or otherwise modifies the signals transmitting programs broadcast by one or more television and AM and FM radio stations and distributes such signals by wire or cable to subscribing members of the public who pay for such services; (2) distributes by cable or wire, news, weather and other information including Civil Defense type information as required as an incidental part of CATV service to all subscribers without additional charge; (3) distributes any and all other lawful communications of a specialized nature provided that it shall not mean or include any facility to which is transmitted any special television program or event for which a separate and distinct charge is made to the subscriber in a manner commonly known and referred to as pay television; except as may be permitted by the Federal, State, and/or local regulatory agencies.

I. *Person* — shall mean any person, firm, partnership, association, corporation, company or organization of any kind.

J. *Applicant* — shall mean any person submitting an application to the City of Louisville for a franchise to operate a CATV system under the terms and conditions set forth by the Board of Aldermen.

K. *Grantee* — shall mean the person to whom or to which a franchise as hereinbefore defined is granted by the Board of Aldermen under this ordinance or anyone who succeeds the person in accordance with the provisions of the franchise.

L. *Gross Receipts* — shall mean all revenue derived directly or indirectly by a Grantee, its affiliates, subsidiaries, parents, and any person in which a Grantee has a financial interest from or in connection with the operation of the CATV system in Louisville, Kentucky, with no deductions whatsoever.

M. *Street* — shall mean the surface of and the space above and below any public street, road, highway, freeway, lane, path, public way, or place, alleycourt, boulevard, parkway, drive or other easement now or hereafter held by the City for the purpose of public travel and shall include other easements or rights of way as shall be now held or hereafter held by the City which shall, within their proper use and meaning entitle the City and its Grantee to the use thereof for the purposes of installing or transmitting CATV System transmissions over poles, wires, cables, conductors, ducts, conduits, vaults, manholes, amplifiers, appliances, attachments, and other property as may be ordinarily necessary and pertinent to a CATV System.

N. *Residential Subscriber* — shall mean a purchaser of any service delivered over the system to an individual dwelling unit where the service is not to be utilized in connection with a business, trade, or profession.

O. *Commercial Subscriber* — shall mean a purchaser of any service delivered over the system who or which is not a residential subscriber.

P. *Basic Service* — shall mean the simultaneous delivery by the company to television receivers (or any other suitable types of audio-video communication receivers) of all subscribers in the City of all signals of over-the-air television broadcasters required by the FCC to be carried by a CATV System as defined hereinabove. Basic service shall also include Grantee channels, City channels, except as may be designated for special purposes by the Mayor, Public channels and Additional channels at the option of the company; or as directed by the Board of Aldermen.

Q. *Additional Service* — shall mean any communications service other than basic service provided over its CATV System by a Grantee directly or as a carrier for its subsidiaries, affiliates, or any other person engaged in communications services including, but not limited to, burglar alarm, data, or other electronic intelligence transmission, facsimile reproduction, meter reading, and home shopping.

R. *Channel* — shall mean a band of frequencies, six megahertz wide in the electro-magnetic spectrum which is capable of carrying either one audio-video television signal and a number of non-video signals or several thousand non-video signals.

S. *City Channels* — shall mean channels on the CATV System which are reserved by this ordinance for use by the City.

T. *Public Channels* — shall mean channels on the CATV System which are reserved for carriage of program material provided by persons who lease channel time and if necessary, studio facilities, from a Grantee for the presentation of programs.

U. *Grantee Channels* — shall mean the channels on the system which are reserved by this ordinance for the carriage of program material originated by a Grantee and the retransmission of broadcast signals in accordance with the FCC's cable casting rules and regulations.

V. *Federal Communication Commission or FCC* — shall mean that agency as presently constituted by the U.S. Congress or any successor agency.

W. *Certificate of Compliance* — shall mean that approval required by the FCC in order for a Grantee of a CATV franchise to begin operation within the City.

X. *Pay Television* — shall mean the delivery over the CATV system of video signals in intelligible form to Residential Subscribers for a fee or charge (over and above the charge for Basic Service) on a per program, per channel or other subscription basis.

To these might be added a definition of "system," as an alternative to CATV defined above, an alternative definition of "gross receipts," and a definition of a "converter," all drawn for illustrative purposes from the New York franchises:

"System" means the broadband communications facility which is to be constructed, operated and maintained by the Company pursuant to this contract.

"Gross Receipts" means all revenue derived directly or indirectly by the Company, its affiliates, subsidiaries, parents, and any person in which the Company has a financial interest, from or in connection with the operation of the System pursuant to this contract, excluding, however, revenues derived from provision of a separate service which uses the System for transmission but including an amount equivalent to what an outside party would have paid for such transmission.

"Converter" means an electronic device which converts signals to a frequency not susceptible to interference within the television receiver of a subscriber, and by an appropriate channel selector also permits a subscriber to view all signals delivered at designated dial locations.

Because subscriber rates and services are generally specified by "dwelling unit," it is also advisable to define this term. One illustration comes from a franchise noted by the Cable Television Information Service:[1]

"Dwelling Unit" shall mean a room or suite of rooms in a building or portion thereof, used for living purposes by one family. A "dwelling unit" shall not mean a building used solely for commercial uses or a Guest House, Guest Room, Hotel or Lodging House.

As will be discussed later, a distinction between single and multiple dwellings for rate purposes often will be useful.

---

[1] Cable Television Information Center, *An Annotated Outline of an Ordinance for Use in Considering a Process for Local Regulation of Cable Television,* p. 19; hereafter cited as *Annotated Outline.*

DURATION OF FRANCHISE

The FCC has substantially restricted the freedom of local authorities to set the duration of the franchise. In its *Report and Order* it specified that:

We are requiring in § 76.31(a)(3) that franchising authorities place reasonable limits on the duration of franchises. Long terms have generally been found unsatisfactory by State and local regulatory authorities, and are an invitation to obsolescence in light of the momentum of cable technology. We believe that in most cases a franchise should not exceed 15 years and that renewal periods be of reasonable duration. We recognize that decisions of local franchising authorities may vary in particular circumstances. For instance, an applicant's proposal to wire inner-city areas without charge or at reduced rates might call for a longer franchise. On the other hand, we note that there is some support for franchise periods of less than 15 years.[2]

However, in its more recent *Reconsideration* it promulgated a stronger rule that "the initial franchise period *shall not exceed* [emphasis supplied] fifteen (15) years, and any renewal franchise period shall be of reasonable duration ..."[3]

Two questions remain with respect to this rule. First, would a longer franchise be permitted if reopener provisions were included after some period of time, e.g., ten years? This approach is employed in the twenty-year New York City franchises, which after ten years are subject to renegotiation of any condition except the duration of the franchise itself:

The franchise shall commence on the effective date of this contract and continue for a period of twenty (20) years, unless sooner terminated as herein provided. However, at any one time after ten (10) years from the effective date, the Board may, upon a review of all the circumstances then affecting broadband communications in the District, notify the Company of its determination that any of the terms and conditions contained herein (except the duration hereof) should be renegotiated, and the Company shall negotiate in good faith with the Board's representatives as to all such terms and conditions. In the event that all such terms and conditions are not renegotiated to the satisfaction of the Board within six (6) months of its notification to renegotiate, the Board may submit any such unresolved matters to arbitration pursuant to Section 20 for a determination consistent with both the public interest and fairness to the Company. The Board's right to initiate renegotiation pursuant to this subdivision shall be cumulative and shall be in addition to and not in derogation of all other rights reserved to the City, the Board and all agencies and officials of the City under other provisions of this contract.

For new franchises not covered by the FCC's grandfathering provisions, this approach probably would not be consistent with the new FCC rules. First, the franchise explicitly makes renegotiation discretionary with the franchising authority, by providing that it "may" decide to reopen the franchise. Second, and more important, the franchise does not provide for modification of the duration, thus making any change—short of franchise revocation—impossible for twenty years.

If drafted somewhat differently, however, this type of approach might be acceptable to the FCC. Provisions for renegotiation of the whole franchise—including its duration—and requiring some form of "public proceeding" might well be considered a bona fide alternative to a shorter initial franchise. In this case, the renegotiation might reasonably be construed as the equivalent of a renewal.

A second question is whether a franchise of indefinite length would be acceptable if it provided for review at relatively short regular intervals. This approach has been recommended by the American Civil Liberties Union:

---

[2] *Cable Television Report and Order* at 3276.

[3] 47 C.F.R. § 76.31(a)(3) (1972), printed in *Reconsideration Opinion and Order,* 37 Fed. Reg. at 13866.

The franchise term shall be unspecified. The franchisor shall review the performance of each of its franchisees at public hearings held no less frequently than bi-annually. All those who wish to present evidence of any kind shall be heard at the hearings. At the conclusion of the hearings, the franchisor must do one of the following: extend the franchise until the next public hearing held under this section; fine the franchisee under the provisions of Section 3.26 [of this franchise] and at the same time set a date for a public hearing under this section, to commence within one year; revoke or cancel the franchise under the provisions of Sections 3.25 and 3.27.[4]

This approach presumably would have a better chance of surviving the FCC's rules, since it provides for total reviews at frequent and fixed intervals. Nevertheless, this approach may be unwise for other reasons. First, it may hamper the franchising process by reducing the ability of prospective applicants to raise venture capital, since neither operators nor investors may be willing to bear the increased risk. Second, even if enforced along lines that will not discourage investment, this approach may result in a de facto permanent franchise—like the virtually automatic renewal of broadcasting licenses.

In any event, the term of the franchise may be important not so much as a device for transferring ownership of the system or building a new one, but rather as a convenient point for reviewing the performance of the cable operator and for renegotiating on the basis of past experience. As an earlier Rand report noted:

> ... The franchise renewal process is [unlikely] to lead to a change in owner-
> ship. In a franchise proceeding, the existing holder has an advantage over
> challengers. In this respect, we should recall that although broadcast li-
> censes are subject to renewal every 3 years, it is a rare occurrence when the
> existing owner loses his license. Even if the existing cable owner was forced
> out, he would be paid some "fair" market value determined perhaps by an
> arbitration board. If the cable operator had performed badly, this fair mar-
> ket value might not cover all debt claims with a reasonable return to equity.
> However, in the case of loss, the underlying difficulty is not that the fran-
> chise is written for, say, only 10 rather than 20 years, but that the cable
> operator has not done well in designing or operating the system or that the
> market is simply not sufficient to permit him to cover costs under *any*
> circumstances.
>
> If it is true that a forced change of ownership is not likely, then why have
> a franchise renewal process at all? The renewal process is useful in at least
> two ways: (1) It provides a formal process for reviewing the performance of
> the operator, and (2) it facilitates renegotiating basic features of the fran-
> chise in accordance with the experience accumulated by the cable operator
> during the preceding period. The process of review assures that the level and
> nature of consumer complaints, growth of the system during the previous
> period, rates charged to subscribers and to other channel users, technical
> standards of service, and other elements can be examined in a more formal
> way than is likely to take place during the franchise period itself. Compari-
> sons between the performance of the cable operator and that of operators in
> other cities would be useful. Although some review will (or should) be con-
> ducted continuously during the operation of the system, the renewal proced-
> ure provides a convenient formal review during which all interested parties
> can come together.
>
> The renewal process also affords the possibility of substantially changing
> the conditions of franchise on the basis of past experience—for example, a
> new set of technical standards based on technological advances that took

---

[4] Jerrold Oppenheim, *Model Code of Regulations: Cable Television - Broadband Communications,* American Civil Liberties Union, Illinois Division, June 1971; hereafter cited as the ACLU Model Code.

place during the earlier franchise period, a modified or new set of fees to be paid to the city, revised procedures by which channels are to be made available to various classes of users, or modification in geographical boundaries of service.[5]

The Illinois Commerce Commission recently observed that:

> There is reason to believe that the emphasis on franchise term may be somewhat misplaced. If a cable system is required at the outset to install adequate capacity and to add to or improve that capacity as the state of the art and market demand develop, and if the quality of the service is subject to adequate supervision, there would seem to be little reason to change or to threaten to change cable operators. If on the other hand a system is under-engineered and under-financed, it may prove difficult to attract another operator to come in and redo the whole system.
> A municipality might however regard a limited franchise term as a useful device to ensure appropriate attention to local needs and desires going beyond the minimum standards that will be imposed by the Commission. This is entirely justified, and the Commission is therefore disposed to regard franchise duration as a local matter to be settled by municipalities subject to FCC guidelines.[6]

While the Illinois Commerce Commission's approach has merit, it assumes a large "if"—namely, that the initial franchise will protect the public interest adequately. Moreover, a franchise renewal proceeding—like a broadcast license renewal proceeding—provides exactly the opportunity for the public input necessary to insure "adequate supervision."

Finally, franchising authorities should consider using a system of fines and forfeitures either in combination with or as an alternative to franchise renewal proceedings. Imposition of monetary penalties through simplified adjudicatory procedures may well be the most workable enforcement method, examples of which are described later. The threat of revoking or not renewing a franchise obviously is more severe, but its very severity probably insures that it will seldom if ever be used. Though the FCC has similar powers over broadcast licenses, it hardly ever has used them.

GEOGRAPHIC EXCLUSIVITY

Most franchises are nonexclusive in the sense that the city reserves the right to franchise more than one operator within the same geographic area.[7] For example, the New York City franchises specify that:

> Nothing in this contract shall affect the right of the City to grant to any other person a franchise or right to occupy and use the streets or any part thereof for the construction, operation, and maintenance

[5] L. L. Johnson et al., *Cable Communications in the Dayton Miami Valley,* The Rand Corporation, R-943-KF/FF, January 1972, Paper 9, pp. 8-9.

[6] Illinois Commerce Commission, *Notice of Inquiry and of Proposed Rule Making—Broadband Cable Communications,* n.d. (early 1972), p. 14; hereafter cited as *Notice of Inquiry.*

[7] In a survey of franchises in New Jersey, 8 out of 66 were found to be exclusive. Leone and Powell, op. cit., p. 258.

of a broadband communications facility within the District or elsewhere, and the Company shall not take a legal position contesting the Board's right to authorize such use of the streets or any part thereof; provided, however, that nothing contained in this subdivision shall prohibit the Company from appearing before the Board and being heard on any application for the grant of such right.

Though New York City adopted nonexclusive franchises partly because of policy considerations and partly because of public pressure, it actually had little choice; the State Constitution bars local governments from granting exclusive franchises—a factor that will vary from jurisdiction to jurisdiction.

Beverly Hills is one of the exceptions to the general practice. It has specified that:

> Pursuant to the provisions of said Chapter 5 of Title 7 of the Beverly Hills Municipal Code, an exclusive franchise to construct, operate and maintain a CATV system within the entire boundaries of the City of Beverly Hills as it now exists or hereafter may be amended, as required by Section 7-5.21 of the Beverly Hills Municipal Code, for a term of fifteen (15) years from the date of acceptance, is hereby granted . . . .

The ACLU Model Code also recommends that "each franchise shall be geographically exclusive."

In general, the practice of granting only nonexclusive franchises has merit. In the words of an earlier Rand report:

> As a practical matter, there is not a great deal of difference between the two types of franchises. In the former case [exclusive franchise], the operator has a *de jure* monopoly. In the latter case [nonexclusive franchise], once he builds a plant he will have a substantial advantage over potential competitors, which gives him a *de facto* monopoly. Our cost analysis does not suggest that it would be economical to have two or more operators with their own lines competing on a house-to-house basis. As in the case of telephone and other public utilities, the construction of duplicate facilities along public rights-of-way would seem wasteful, at least at this stage of cable development. Indeed, it remains an open question whether even a *single* operator can make a profit in large cities having extensively developed over-the-air broadcasting service.
>
> All in all, there is nothing to lose and perhaps something to gain by writing only nonexclusive franchises. If the operator is doing a good job, the threat of additional competition would be inconsequential, and the two types of franchises would have the same effect; but the potential threat of competition under a nonexclusive franchise would provide additional stimulus for the existing operator to perform well. If worst comes to worst and he does a poor job, then competition would serve as a safety valve to protect the public interest.[8]

BROADCAST SIGNALS TO BE CARRIED

On the federal level, most of the sound and fury concerning cable's development has centered on its carriage of broadcast television signals—which broadcasters view as "unfair competition" and which copyright interests see as robbery of their rightful profits. As a result, the FCC has devoted most of its efforts for the last six

---

[8] Johnson et al., op. cit., Paper 9, p. 8.

years to settling the signal carriage issue, and most of the rules adopted by its *Cable Television Report and Order* relate to signal carriage. In brief, the FCC's new rules require cable systems to carry all "local" television stations and usually permit them to carry no more than a few "distant" television stations.[9]

Despite the fact that the FCC has preempted practically all regulation of broadcast signal carriage, a few interstices remain that a local franchising authority may fill. First, where the FCC's rules give a cable system some choice as to the stations from which it draws its "distant signal" complement, the franchising authority and the local community may wish to exercise some influence. Though under the FCC's rules a franchising authority may not specify the signals a cable system will carry, some mechanism for consultation between the system and the franchising authority appears to be permissible.

Second, the FCC's rules allow a cable system to import an unlimited number of educational and foreign language distant signals as long as, with respect to the former, local educational authorities do not object. The franchising authority may wish to have some input into the cable operator's choice and number of these signals.

Finally, all of the FCC's restrictions and requirements are subject to waiver, as noted before in Sec. II. As a result, the franchising authority may encourage the cable system to apply for a waiver, support the cable system's petition for a waiver, or oppose the cable system's petition for a waiver. In all these cases, the FCC presumably will give considerable weight to the desires of the cable system's community, if forcefully expressed. Though all these courses will involve the franchising authority in complex administrative procedures, they nevertheless should be considered. A companion study describes the problems and procedures in more depth.[10]

CONSTRUCTION TIMETABLE

In response to concern that a franchisee may hold onto a franchise, hoping for a profitable resale and meanwhile dragging his feet on construction, the FCC has set down guidelines for construction and operation:

We are establishing in § 76.31(a)(2) general timetables for construction and operation of systems to insure that franchises do not lie fallow or become the subject of trafficking. Specifically, we are providing that the franchise require the cable system to accomplish significant construction within 1 year after the certificate of compliance is issued, and that thereafter energized trunk cable be extended to a substantial percentage of the franchise area each year, the percentage to be determined by the franchising authority. As a general proposition, we believe that energized trunk cable should be extended to at least 20 percent of the franchise area per year, with the extension to begin within 1 year after the Commission issues its certificate of compliance. But we have not established 20 percent as an inflexible figure, recognizing that local circumstances may vary.[11]

The FCC's rules do not require a cable system to wire any specified percentage of their franchise area in any specified amount of time. Though the FCC "believes" that

[9] For a fuller discussion of the rules' intricacies, see the companion study by S. Rivkin, *Cable Television: A Guide to Federal Regulations*, Crane, Russak & Co., New York, 1974.

[10] Monroe Price and Michael Botein, "Citizen Participation After the Franchise," Chapter 5 of this book.

[11] *Cable Television Report and Order* at 3276.

20 percent per year would be a reasonable figure, its requirement of only a "substantial percentage" and its recent actions suggest that it will take a liberal attitude. Moreover, even strict enforcement of a specified percentage requirement would not by itself insure equitable wiring of a franchise area, since it would allow a cable operator to wire more affluent neighborhoods first in order to meet his percentage quota. As a result, a franchise authority should specify not only the amount of the franchise which must be wired each year, but also the geographic distribution of it.

For appropriate wording of provisions consistent with the FCC regulations, the New York City franchises may provide a useful guide:

> The Company shall extend the installation of cables, amplifiers and related equipment throughout the District as rapidly as is practicable. Within four (4) years from the effective date of this contract, the Company's trunk line installations of cable, amplifiers and related equipment shall be capable of providing Basic Service to every block within the District. Thereafter, the Board may impose such further construction obligations as are necessary to bring Basic Service to any building within the District.

Alternative wording is provided by a draft model franchise of the Miami Valley Council of Governments (MVCOG):

> The grantee shall construct one head-end and the necessary antenna and studio facility to permit the reception of broadcast signals and the origination of programming within one year after the effective date of this franchise. The grantee further shall complete construction of at least twenty percent (20%) of the cable distribution plant during the first year after the effective date of this franchise and shall during such year commence construction of a separate trunk cable from a head-end facility for each Separate Service Area as provided in subsection 5(a) hereof. Thereafter, the grantee shall complete construction annually of at least twenty percent (20%) of the distribution plant and any remaining uncompleted portion of the cable television system necessary to fulfill the obligations of this franchise or of the F.C.C. regulations.

> Construction of the system shall proceed in a non-discriminatory manner that provides relatively equivalent service to each Separate Service Area and that meets with the approval of the franchisor.[12]

EXTENT OF WIRING IN THE FRANCHISE AREA

Closely related to the question of construction timetables is the issue of what amount of the franchise area is to be wired. FCC regulations clearly prohibit the cable operator from skimming the cream off the most profitable portions of the geographic area:

> We emphasize that provision must be made for cable service to develop equitably and reasonably in all parts of the community. A plan that would bring cable only to the more affluent parts of a city, ignoring the poorer areas, simply could not stand. No broadcast signals would be authorized under such circumstances. While it is obvious that a franchisee cannot build everywhere at once within a designated franchise area, provision must be made that he develop service reasonably and equitably. There are a variety of ways to divide up communities; the matter is one for local judgment.[13]

Aside from the question of cream-skimming, there is the important issue of what percentage of homes in the franchise area are able to have access to the cable system. Must the cable plant pass within the normal distance of *every* home, regardless of expense? As noted in an earlier Rand report:

[12] Miami Valley Council of Governments, CATV Subcommittee, *Proposed Model CATV Franchise,* July 1972; hereafter cited as the MVCOG Model Franchise.
[13] *Cable Television Report and Order* at 3276.

The major problem with insisting on literally 100-percent coverage is that in nearly any large franchise area a few homes will be extraordinarily expensive to wire because of geographical locations that require additional expensive trunk and feeder lines to maintain good signal quality. Other expenses are incurred where there is a sudden fall-off in population density in a small subarea; this generates a very high cost for those few additional homes passed by the cable. Our financial projections for the Dayton area suggest that the average cost of cable plant per home passed is about $120. For a few homes, however, this cost could run to two or three times as much (the precise figure cannot be determined until a detailed street-by-street engineering blueprint is drawn—a task normally done by the cable operator shortly before he commences installing cable in a given subarea of his franchise.) A major policy question is whether other subscribers should bear the cost burden of the abnormal difficulty of wiring these few homes. This problem is especially worrisome since it is likely that the homes that are the least difficult to wire will be located in the densely populated low-income areas, while the few homes with geographical wiring problems are likely to be in the high-income suburbs.[14]

One possibility in coping with this issue is to follow New York City's lead and require that *all blocks* of the city have access, leaving for subsequent determination the extent to which service will be offered to each separate building or residence in the block.

Another possibility is a provision not requiring every block to be wired, but instead enabling a special charge to be levied for homes in particularly difficult blocks. The provision for cable extension charges in the Beverly Hills franchise may prove a useful guide:

In the event that a potential subscriber's premises are located at such a distance from the feeder cable that it is not economically feasible for the Grantee to provide service at the foregoing rates and charges, the City Council shall determine, upon request from the potential subscriber or the Grantee, the amount, terms, conditions and refund provisions of the line extension charge which, in addition to the foregoing rates and charges, would be fair and reasonable under the particular conditions and circumstances.

One major problem here, however, is that the provision begs the question posed earlier; the "distance from the feeder cable" itself depends on how much cable plant the operator builds in the first instance. If he is expected to pass closely only 70 percent of the homes, then many homes will be subject to special charges under this provision. If he is expected to pass 100 percent, then by definition the provision would never apply. A second problem with this approach is that it does not provide any standards for determining the "amount, terms, conditions and refund provisions of the line extension charge . . ." Similarly, requiring an administrative proceeding in order to settle the issue might generate excessive and unnecessary transactional costs.

The franchise should require the cable operator to build plant passing within 150 feet (or some other "normal" distance) of perhaps 90 or 95 percent of the homes and *in addition* include a provision (perhaps along the lines of the Beverly Hills approach) to take care of the remaining 5 or 10 percent, but with well-defined conditions under which service is to be extended.

[14] Johnson et al., op. cit., Paper 9, pp. 12-13.

CONSTRUCTION REQUIREMENTS

Underground Versus Aboveground Construction

Several provisions are advisable with respect to underground versus aboveground construction, standards for conduit construction, condition of streets, and other factors. A cable operator generally is expected to rely on aboveground construction by using existing telephone and power poles whenever available, and is expected to use underground construction where existing utilities are already underground. One example of useful language is a franchise provision noted by the Cable Television Information Service:

> In areas of the city having telephone lines and electric utility lines underground, whether required by ordinance or not, all or any CATV permittees' lines, cables and wires shall be underground. It shall be the policy of the city that existing poles for electric and communication purposes be utilized wherever possible, and that underground installation even when not required is preferable to the placing of additional poles.[15]

Although the city's "policy" here may be satisfactory, a more practical and detailed enforcement mechanism is desirable. The Beverly Hills franchise provides a good example:

> Poles shall not be installed for the sole purpose of supporting CATV cable without written justification and approval by the City Engineer... The City Engineer shall require street crossings to be placed underground if there are no other overhead wires at the crossing.

This type of condition seems reasonable on two counts. First, for aesthetic reasons, most observers would agree that the landscape should not be cluttered with poles for the *sole* purpose of stringing coaxial cable, when so much effort already has been devoted to placing other utilities underground. In some cases existing underground duct space is sufficient to carry cable also; but even if this is not true, the cable operator generally would be expected to go underground and bear the full cost of retrenching. In this latter case occasional exceptions might be made by adding language such as, "Poles shall not be installed for the sole purpose of supporting CATV cable without written justification and approval by the City Engineer," as in the Beverly Hills franchise.

Second, to insist on underground cable when *existing* utilities are aboveground would place a very severe cost burden on the cable operator with questionable benefits to the public. Cable strung on existing poles would add little to the aboveground jungle of wires; hence, the aesthetic benefits in this case would be small. Moreover, the costs of underground construction can be two, three, or many times as high as the cost of aboveground construction, depending on soil and other conditions. Aboveground construction, including cable and amplifiers, generally costs from $4000 to $8500 per mile, depending on the kind of cable plant to be built; underground construction can vary from $10,000 per mile in favorably located areas to as much as $100,000 per mile or even more in New York City.[16] Overhead housedrops also are obviously less expensive than underground connections.

---

[15] Cable Television Information Center, *How to Plan an Ordinance*, p. 23.

[16] For a discussion of the wide variations in underground construction, see W. S. Comanor and B. M. Mitchell, *The Economic Consequences of the Proposed FCC Regulations on the CATV Industry*, The National Cable Television Association, Inc., Washington, D.C., 1970, Appendix C.

Yet some franchises require underground construction irrespective of the nature of existing utilities. The San Mateo franchise flatly states that "all construction shall be underground." The Sunnyvale franchise also originally included such a provision. The cable operator there ran into such high costs in attempting to comply, however, that he eventually succeeded in getting permission to use existing utility poles in areas yet to be wired.

Standards for Underground and Aboveground Construction

Typically, franchises also include provisions regarding construction standards and practices for both aboveground and underground construction. For aboveground construction, an example of useful language is drawn from a franchise noted by the Cable Television Information Center:

> Each CATV permittee's distribution system in the public streets shall comply with all applicable laws, regulations, and ordinances, and all its wires and cables suspended from poles in the streets shall comply with the minimum clearances above ground required for telephone lines, cables, wires and conduits.[17]

Though this provision seems deceptively simple on its face, it covers a multitude of situations and at the same time maintains the community's aesthetic balance merely through incorporating existing standards by reference.

A good example of underground standards is included in the relatively elaborate franchise of the city of Beverly Hills, which has paid special attention to underground construction standards because other utilities are underground in large portions of the city.

> All underground cable on public rights-of-way and service laterals on private property shall be placed in conduits. Property owners requesting exception from this requirement must do so in written form to the City Engineer.
>
> Conduit material shall be in accordance with recognized industry-wide standards as approved by the City Engineer. Any material authorized must have been in widespread use for a minimum period of three years.
>
> CONDUIT INSTALLATION—All conduits shall be placed by approved boring or jacking methods with a minimum cover of 24 inches, except for such separate specific authorization by the City Engineer due to unusual circumstances. Permission will not be granted to cut sidewalks, driveways, streets, alleys and parkways for continuous trench to lay conduit with the following exceptions:
>
> (1) Permission will be granted to trench around obstructions and substructures when necessary and for trench pits incident to jacking and boring operations.
> (2) In the event it can be proved conclusively that conduit cannot be installed by jacking or boring operation, permission will be granted to install conduit by trenching methods.

Safety and Damage Requirements

Many franchises contain provisions to protect the public from harm and undue inconvenience, to ensure that any physical damage is properly repaired by the cable operator, and to put him on notice that he may have to modify his plant as a consequence of physical changes in the community. The Akron franchise states that:

[17] *An Annotated Outline,* p. 41.

The Company's transmission and distribution system, poles, wires, and appurtenances shall be located, erected and maintained so as not to endanger or interfere with the lives of persons, or to interfere with improvements the City may deem proper to make, or to hinder or obstruct the free use of the streets, alleys, bridges, or other public property. Removal of poles or equipment when necessary to avoid such interference will be at the Company's expense.

In the maintenance and operation of its television transmission and distribution system in the streets, alleys, and other public places, and in the course of any new construction or addition to its facilities, the Company shall proceed so as to cause the least possible inconvenience to the general public; and any opening or obstruction in the streets or other public places made by the Company in the course of its operations shall be in accordance with the Rules and Regulations Governing the Making of Openings in Streets, Sidewalks, Public Ways or Places of the City of Akron, Ohio, as established by the Department of Public Service of said City, and which are in effect at that time.

The model ordinance of the National Institute of Municipal Law Officers (NIMLO) states that:

In case of disturbance of any street, sidewalk, alley, public way, or paved area, the grantee shall, at its own cost and expense and in a manner approved by the [Director of Public Works or other appropriate official], replace and restore such street, sidewalk, alley, public way, or paved area in as good a condition as before the work involving such disturbance was done.

If at any time during the period of this Franchise the City shall lawfully elect to alter or change the grade of any street, sidewalk, alley, or other public way, the grantee, upon reasonable notice by the City, shall remove, relay, and relocate its poles, wires, cables, underground conduits, manholes, and other fixtures at its own expense.

Any poles or other fixture placed in any public way by the licensee shall be placed in such manner as not to interfere with the usual travel on such public way.

The grantee shall, on the request of any person holding a building moving permit issued by the City, temporarily raise or lower its wires to permit the moving of buildings. The expense of such temporary removal or raising or lowering of wires shall be paid by the person requesting the same, and the grantee shall have the authority to require such payment in advance. The grantee shall be given not less than forty-eight (48) hours' advance notice to arrange for such temporary wire changes.

The grantee shall have the authority to trim trees upon and overhanging streets, alleys, sidewalks, and public ways and places of the City so as to prevent the branches of such trees from coming in contact with the wires and cables of the grantee, except that at the option of the City, such trimming may be done by it or under its supervision and direction at the expense of the grantee.[18]

One interesting aspect of this provision is that by authorizing the cable operator to use private and public property, the city actually is giving him a de facto right of eminent domain—a practice that may be questionable under some states' laws. This approach thus is similar to authorizing cable operators to wire apartment houses without their owners' consent, as will be discussed later.

As another example of similar provisions, the New York franchises specify that:

No construction, reconstruction or relocation of the System, or any part thereof, within the streets shall be commenced until written permits have been obtained from the proper City officials. In any permit so issued, such officials may impose such conditions and regulations as a condition of the granting of the same as are necessary for the purpose of protecting any structures in the streets and for the proper restoration of such streets and structures, and for the protection of the public and the continuity of pedestrian and vehicular traffic.

Should the grades or lines of the streets which the Company is hereby authorized to use and occupy be changed at any time during the term of this contract, the Company shall, if necessary, at its own cost and expense, relocate or change its System so as to conform with such new grades or lines.

[18] Robert L. Winters, *Municipal Regulation of "CATV"—Community Antenna Television—Model Ordinance*, National Institute of Municipal Law Officers, Washington, D.C., 1967; hereafter cited as the NIMLO Model Ordinance.

Any alteration to the water mains, sewerage or drainage system or to any other municipal structures in the streets required on account of the presence of the System in the streets shall be made at the sole cost and expense of the Company. During any work of constructing, operating or maintaining of the System, the Company shall also, at its own cost and expense, protect any and all existing structures belonging to the City. All work performed by the Company pursuant to this subdivision shall be done in the manner prescribed by the City officials having jurisdiction therein.

Although the New York City and the NIMLO provisions require the cable operator to protect individuals' rights, question arises as to how individuals are to seek remedy. An outraged resident can complain to the appropriate official, who may then take the necessary action. In addition, individuals could exercise a private legal right, which they could enforce either in an existing small claims court or, perhaps even better, in a specially constituted tribunal established under terms of the franchise.

EMPLOYMENT PRACTICES AND TRAINING

Franchising authorities will need to include adequate provisions to ensure that the cable operator's personnel practices are nondiscriminatory. One example of employment provisions, drawn from the New York City franchises, is noteworthy:

The Company will not refuse to hire or employ, nor bar or discharge from employment, nor discriminate against any person in compensation or in terms, conditions or privileges of employment because of age, race, creed, color, national origin or sex.

The ACLU Model Code goes further in requiring that the cable operator actively seek out and train minority-group employees at a level reflecting their percentage of the franchise area's total population:

No franchise shall discriminate on the basis of sex, race, national origin, religion, creed, or arrest or conviction records in hiring and promoting employees. Each franchisee shall seek out and train employees so that minority groups are represented in its employee work force in the same relative proportion as they are represented in the population of the franchise area. Each grantee shall file an affirmative action plan to this end annually with the *Commission.* This plan shall include a report of persons employed, together with their positions and salaries, by categories listed in the first sentence of this section.[19]

However, no matter how well intentioned this type of provision, it creates problems of which the franchising authority may not be initially aware. First, hiring requirements must mesh with existing bans on discrimination; these may include local or state laws and definitely will include the FCC's antidiscrimination rules, the Civil Rights Act of 1964, and the equal protection clause of the federal Constitution. Second, the experience with bans on age and sex discrimination has been extremely frustrating; the line between a legitimate classification and an invalid discrimination is highly tenuous. Third, attempts to define that action which is affirmative often end up in either tokenism or quotas, the former of which is inadequate and the latter of which promotes backlash. The courts have had the bitter experience

[19] Here and elsewhere the ACLU distinguishes between the "Commission" and the "franchisor." The franchisor, such as a city council, grants the franchise, while some other agency such as a new city commission or a state commerce body is responsible for continuing regulation.

of finding that attempts to remedy past violations of the equal protection clause often become present violations of the equal protection clause. The franchising authority should move carefully in drafting these provisions in a manner that is both practical in enforcement and consistent with applicable laws.

TECHNICAL STANDARDS

The many questions relating to technical standards that concern local officials are treated in a separate Rand study. This report deals with some of the more general language relating to operating standards and to those modifications necessary to reflect technical advances.

One of the most serious shortcomings of many franchises is their vague language regarding technical and operational standards; this flaw can lead to serious disagreement and debate. In some cases, a literal interpretation of the language would impose exorbitant costs on the cable operator that one way or another would have to be passed on to the public. For example, one franchise quoted by the Cable Television Information Center states:

> The CATV system shall be installed and maintained in accordance with the highest and best accepted standards of the CATV industry, to the effect that subscribers shall receive the highest quality service technically possible.[20]

One obvious problem here is how to interpret the phrase "highest and best accepted standards." The standards of the cable television industry have many dimensions; what is well accepted by some areas or experts is disputed by others. Cable is still an infant industry and much disagreement remains about how cable plant should be built (because of the many design and construction alternatives as well as different areas' requirements) and how the many tradeoffs in design and cost should be taken into account.

No less troublesome is the above franchise's requirement that the "subscribers shall receive the highest quality service technically possible." A chasm divides the *technically possible* from the *economically feasible*. Without question, cable technology could extend superlative service to every subscriber by spending money freely—on oversize cable, more than the usual number of headends, closely spaced amplifiers, cable plant able to withstand the severest wind storms, large maintenance crews to keep the system in excellent repair (including adjustment of amplifiers for day-to-day temperature variations), and a large office force to handle customer complaints and queries. But such lavishness could easily double or triple construction and operation costs—costs that one way or another must be borne by the user.

The same franchise goes on to state:

> The System, in addition to meeting the standards herein set forth, shall continue to conform to the highest state of the art in the field of Community Antenna Television and shall continue to be designed, redesigned, installed, operated and equipment replaced and maintained in accordance with the best engineering practices in the industry.

[20] *An Annotated Outline*, p. 46.

Again, experts disagree sharply about what constitutes the best engineering practice in particular circumstances. Also, continuing to upgrade the system in accordance with the "best" practices again could lead to extraordinarily high costs. In redesigning an existing system, the franchising authority clearly must consider the usefulness of the existing system to the public as well as the cost/benefit relation of redesigning and reconstructing the plant immediately or gradually. As an example, it hardly would be in the public interest to require the telephone industry to switch overnight to the newest and most modern telephone instruments. Nor would it have been in the public interest to force the airline industry to abandon propeller aircraft immediately upon the advent of jets.

In view of the preceding examples, why then do cable operators willingly agree to such franchise language? The answer is simply that the language is *not* interpreted literally; as a practical matter, the cable operator has leeway in designing and maintaining a system with inevitable variations in service quality.

Nonetheless, it makes little sense for a franchise to embody language that is unenforceable and that could create serious mischief in interpretation. A better approach is to specify technical and operational standards (including adequate margins to recognize that perfection costs money) and to establish a schedule of well-defined penalties and arbitration procedures for failure to perform adequately. Combined with such specific provisions, a general "state of the art" clause may have residual value by reserving power for the franchising authority to cope with new technological developments—which always are much clearer by hindsight than by foresight.

OPERATIONAL STANDARDS

Merely insuring that a system's design is relatively up-to-date, however, does not guarantee subscribers adequate service; a well-designed but poorly maintained system obviously leaves much to be desired. A franchising authority must therefore insure that the cable operator delivers the promised service. Indeed, the FCC's new regulations explicitly provide that franchising authorities must be receptive to subscriber complaints, by requiring that:

> The franchise shall specify procedures for the investigation and resolution of all complaints regarding the quality of service, equipment malfunctions, and similar matters, and shall require that the franchisee maintain a local business office or agent for these purposes.[21]

Thus, the franchisor must determine what enforcement mechanisms or penalties it will invoke when citizens register valid complaints against the cable system. Two principal approaches have been used. One is to define operating standards in a general way in the franchise and negotiate with the cable system for corrective measures when necessary. For example, the New York City franchises for Manhattan state:

---

[21] 47 C.F.R. § 76.31(a)(5) (1972), printed in *Cable Television Report and Order* at 3281.

The Company shall furnish to its subscribers and customers for all services the best possible signals available under the circumstances existing at the time, to the satisfaction of the Director of Communications ...

Other cities have used similar language.

In principle, the advantages of this approach are that a city retains great flexibility in handling complaints and can negotiate with the operator from a position of strength, since it has final authority to terminate or not renew his franchise. However, several problems may arise with this approach:

- There are no objective standards for determining the validity of subscriber complaints;
- A city must decide in each instance how many complaints are needed before it should act;
- Citizens have no objective way of measuring the performance of the enforcement agency;
- The city is vulnerable to charges, however unfounded, that it selectively enforces complaints or otherwise deals under the table with the operator;
- The "ultimate sanction" of franchise termination or nonrenewal simply may not be credible in dealing with minor, day-to-day problems.

The second approach would attempt to set specific operating standards and appropriate penalties for noncompliance in the franchise. The franchise might specify a maximum time for the operator to respond to subscriber complaints (for example, same-day response if the complaint is received before noon). It might set standards for system reliability in terms of the number of service interruptions per year, or establish picture quality standards over and above those contained in the FCC rules. For each of these examples, the penalty for noncompliance might be a rebate of part of the monthly fee to each affected subscriber.

An example of such a system of procedures and penalties is drawn from the Illinois Commerce Commission:

System operators will have to provide same-day service response seven days a week for all complaints and requests for adjustments received before 2:00 p.m. each day. Calls received after 2:00 p.m. must be responded to within 24 hours.

If the operator fails to remedy a loss of service attributable to the cable system within 24 hours after a complaint, he will be required to rebate one-thirtieth (1/30) of the regular monthly charge to each subscriber for each 24 hours or fraction thereof following the first 24 hours after a report of loss of service, except to the extent that restoration of service is prevented by strike, injunction, or other cause beyond the operator's control.

The operator will have to keep a log, and file a copy thereof with the Commission at the close of each calendar quarter, listing by category all complaints and trouble calls received, the number of second or subsequent calls on the same complaint, the remedial action taken, the period of time required to satisfy each reported complaint, and the rebate (if any) to subscribers.[22]

The MVCOG model franchise also contains reasonably clear language about required performance and penalties for nonperformance:

Whenever it is necessary to interrupt service for the purpose of making repairs, adjustments, or installations, the grantee shall do so at such time as will cause the least amount of inconvenience to the

---

[22] *Notice of Inquiry*, p. 29.

subscribers; and, unless such interruption is unforeseen and immediately necessary, it shall give reasonable notice thereof to its subscribers.

Whenever the grantee's system shall be out of service so that the subscribers may not receive more than half of the available channels for a period of 48 hours, then the grantee shall automatically refund to the subscribers a proportionate amount of his subscription fee.

The requirements for maintenance of equipment contained in this provision shall not apply to the subscriber's television receiver.

It is relatively easy to enforce specific standards in an open and straightforward manner. The disadvantages of the approach are that the franchising authority and the operator may be unable to agree on operational standards, and that the city has less flexibility to take special factors into account in enforcement. In essence, this approach is like handing out tickets for minor traffic offenses, rather than negotiating with the violator to improve his behavior under threat of revoking his driver's license. And while states do revoke drivers' licenses, there is no evidence to date that municipalities will revoke or refuse to renew franchises.

Perhaps the best solution is a combination of the two approaches. Operational performance standards might be included in the franchise where they can be reasonably determined, with appropriate penalties for noncompliance. Where setting operational standards is difficult—as, for example, the picure quality of locally originated programming from remote sites—the city must rely on negotiating any needed improvements.

Finally, any effective complaints system must include workable remedies as well as rights. One possible alternative is arbitration, which the New York City franchises provide for:

Matters which are expressly made arbitrable under provisions of this contract shall be determined by a panel of three arbitrators appointed by the Presiding Justice of the Appellate Division of the Supreme Court of the State of New York for the First Judicial Department. The fees of the arbitrators shall be fixed by the said Presiding Justice. The expenses of the arbitration, including the fees of the arbitrators, shall be borne by the parties in such manner as the arbitrators provide in their award, but in no event will the City be obligated for more than half the expenses. The determination of a majority of the arbitrators shall be binding on the parties. In the event that an arbitrable matter arises contemporaneously under another franchise, involving the same issue as that to be arbitrated under this franchise, the Company will not claim or assert that it is prejudiced by, or otherwise seek to prevent or hinder, the presentation of the arbitrable matter under such other franchise for determination by a single panel.

As an alternative to relying on the courts in selection of arbitrators, franchising authorities should consider following the standard American Arbitration Association practice of allowing each party to designate one arbitrator and then having the two arbitrators pick a third.

ACCESS TO PREMISES BY CABLE OPERATOR

One of the major difficulties in some large urban areas is that many landlords of apartment buildings refuse to allow the cable operator to wire individual apartments at the tenant's request. Some landlords charge tenants for using their own master antenna systems and do not want the cable operator to compete with this service. Others are apprehensive about physical disruption and possible damage to the building resulting from installation of cable in cramped or difficult quarters. And many landlords are simply out to get what they can; in New York City some have been known to demand payments from the cable operator for permission to wire the building.

Franchise authorities therefore should consider provisions that both assure access by the cable operator and protect the landlord from property damage. The provisions of the ACLU Model Code may serve as an appropriate guide:

> Each franchisee shall have the right and obligation to provide cable television service to any member of the public in any publicly or privately owned buildings that are in the franchisee's franchise area without paying a charge to the building owners. Any disputes between the franchisee and any building owner shall be heard at and resolved by a public hearing by the franchisor. Each franchisee shall report to the franchisor any building owner who requests a payment from the franchisee before allowing the franchisee to install cable system service in the building or who otherwise refuses it free access. The franchisor after public hearing, may fine such building owners up to $ ___ per offense and order the building owner to allow the franchisee access. If a building owner is for any reason not available to appear before the franchisor, the franchisor may proceed against any agent of the building owner.

> Any damage caused to the property of building owners or users or any other person by the grantee shall be repaired fully by the franchisee.

This type of franchise provision is by no means self-executing. Terms such as "free access" and "repaired fully" are vague and may create disputes. Moreover, some local governments may lack the power to impose such requirements on landlords and therefore must seek it from the state legislature. As an alternative, a franchising authority might require the cable operator to pay landlords a flat fee for each apartment passed, in order to avoid the delays and costs of litigation. Or the franchising authority might specify a flat fee but allow dissatisfied landlords to commence proceedings for a higher fee; in these proceedings the landlord would have the burden of proof or even the burden of rebutting a presumption that the flat fee is reasonable.

RATES AND OTHER CHARGES TO THE SUBSCRIBER: GENERAL CONSIDERATIONS

Within the new FCC rules, one area that remains open for local determination is the matter of subscriber rates and other charges. In the opinion accompanying its rules, the FCC specifies:

> We are permitting local authorities to regulate rates for services regularly furnished to all subscribers. The appropriate standard here is the maintenance of rates that are fair to the system and to the subscribing public—a matter that will turn on the facts of each particular case (after appropriate public proceedings affording due process) and the accumulated experience of other cable communities.[23]

Here the Commission apparently attempted to require rate regulation but not specify the type. (Its subsequent actions indicate that it will accept almost any form of rate control.) On the one hand, the FCC apparently felt that some rate control was necessary. On the other, it presumably wanted to avoid traditional public utility rate-of-return regulation, which would inhibit the flow of venture capital into cable television. Moreover, the rule specifically stipulates regulation only of rates "for

---

[23] *Cable Television Report and Order* at 3276.

installation of equipment and regular subscriber services."[24] The question of regulating leased-channel rates for new services—e.g., pay television and information retrieval—is left open.

Thus local authorities have full jurisdiction (subject to possible state preemption) to review and approve fees for cable installations, monthly services, outlet relocations, and so forth. In setting rates, the local franchising authority should consider several points:

• First, it must decide what standards, if any, to use. A number of alternatives are available. First, it might adopt traditional public utility rate-of-return regulation. The problems of estimating the investment on which the rate of return is regulated make this approach difficult.[25] Second, it might use rate surveillance—that is, policing a system's profits only to insure they do not become exorbitant. This approach not only is administratively easy for the franchising authority, but also allows cable systems to earn enough profit to attract venture capital. Finally, the franchising authority might assume a totally "hands-off" attitude and simply approve whatever rates the cable system proposed. Though this might satisfy the letter of the FCC's rules, it obviously would present difficulties in terms of both policy and politics.

• Differences in rates for various services or components of services generally should reflect differences in underlying costs, if the franchising authority seeks to prevent one group of users from heavily subsidizing another. To be sure, the cost of service for any two subscribers is not identical, and rates cannot reflect individual cost-differences. Some rate averaging is inevitable. In most franchises the basic monthly charge is constant over the whole franchise area, despite cost-differences among locations because of factors such as population density, construction conditions, and topography. At the same time, other cost-differences should be taken into account. The subscriber who requires an underground connection should pay more than the one who requires only a conventional overhead dropline. The subscriber who lives more than some "reasonable" distance from the nearest feeder line should pay a surcharge. Conversely, subscribers in densely packed multiple dwelling units should pay less than those in single-family residences.

• As noted above, the FCC's rules presently may bar, without special justification, rate regulation of leased channels for new services. Ample leeway should be built into the franchise, however, to encompass regulation of subscriber rates for new services that today are only on the horizon. The long experience with cable retransmission of broadcast signals and small-scale local program origination provides a reasonably sound basis for establishing basic subscriber rates and connection charges for these conventional uses. But what about services that may be added in the future? These may include: two-way leased lines for data transmission; special channels for pay movies; connection and maintenance of new kinds of terminals that may work in conjunction with or independently of the conventional home television set. No satisfactory basis exists today for setting specific rates for such services—a factor that probably influenced the FCC in not specifying the regulation of these services' rates. The best approach is to make the franchise flexible by reserving the

---

[24] 47 C.F.R. § 76.31(a)(4) (1972), printed in *Cable Television Report and Order* at 3281.

[25] These difficulties will not be recounted in detail here. An extensive review is provided by Posner, *National Monopoly and Its Regulation*, 21 *Stan. L. Rev.* 548 (1969).

right of the local or state agency to establish, review, or modify rates for these services as they arise and as the FCC permits.

• Mechanisms should be built into the franchise to allow necessary rate changes as circumstances warrant. General inflationary pressure is beyond the control of the cable operator and may require periodic rate increases during the franchise term —especially if it is as long as 15 or 20 years. Experience may show that some services are not paying their way and need to be repriced. On the other hand, extraordinary profits would merit rate reduction. As noted at the beginning, the standard that the franchising authority adopts for rate control will influence all of these judgments.

• As a corollary, the franchising authority must take care that in competing for the franchise the winner not be forced into proposing nonremunerative rates. The FCC is concerned about similar problems in other areas of franchising, such as with respect to franchise fees. Since competitors for a franchise tend to offer overly generous franchise fees at the expense of services to the public, the Commission has taken a strong stand against franchise fees it regards as excessive, as will be discussed later. Because competitors also tend to offer large numbers of free channels for various public uses—again constituting a potential burden on other users—the Commission hass banned the wholesale giveaway of channels, as will be discussed later. Yet the possibility of excessive rate competition remains. If franchise applicants cannot compete on the basis of free channels and franchise fees, they may be under much more pressure to compete in terms of rates. Although some rate competition is clearly in the public interest, it creates the danger of a candidate's offering uneconomically low rates. Correctly or incorrectly, he may reason that once he obtains the franchise he can obtain an increase; or he may be overly optimistic about his level of expected costs and the expected growth in number of subscribers. In either case, saving a partially built plant from bankruptcy will require increasing subscriber rates, perhaps after long renegotiation, during which time service to the public is disrupted or poor.

In light of the preceding points, we shall now discuss the separate detailed elements of different subscriber rates. The wording of existing franchises will illustrate approaches to be considered and pitfalls to be avoided.

CABLE CONNECTION FEES

In addition to monthly service fees and other charges, cable operators generally set a one-time fee for connecting the subscriber to the nearest feeder line. Where the feeder line is on telephone or electric power poles, the cable connection is a dropline from the nearest pole. If the feeder line is underground (as in many downtown areas and some new residential areas) a trench is dug from the feeder to the home, as with other underground utilities.

As mentioned earlier, the cost of connection depends heavily on whether underground construction is involved—a factor that should be considered in the franchise, as previously discussed. The subscriber's distance from the feeder line is also an important consideration. An interesting approach to both these points is contained in the Beverly Hills franchise:

I. RATES AND CHARGES FOR INSTALLATION SERVICES IN SINGLE FAMILY AND DUPLEX RESIDENCES.

(a) First Outlet

1.1 Overhead Service

   a. The charge shall be $19.50 for normal installation.

   b. In the event that the distance from the center line of the street, alley, or easement occupied by the feeder cable to any outlet at the subscriber's set exceeds 150 feet, the Grantee may make an additional charge not to exceed the actual direct cost to the Grantee attributable to such distance in excess of 150 feet.

1.2 Underground Service From Underground Feeder Cable

   a. In the event that the subscriber independently provides for his own trenching (including backfilling, repaving and/or replanting), the charge shall be as stated above for Overhead Service—1.1 a. and 1.1 b.

   b. In the event that the subscriber contracts with the Grantee to provide the trenching and related work, the charge for the trenching, et cetera, shall be the variable cost incurred by the Grantee and the charge for the duct, cable and installation shall be as stated above for Overhead Service—1.1 a. and 1.1 b.

   c. In either event, Grantee shall bear the full cost of providing the trenching and all other facilities from the feeder cable to the subscriber's property line.

1.3 Underground Service From Overhead Feeder Cable

The charge shall be as stated above for Overhead Service—1.1 a. and 1.1 b., and in addition thereto, the difference between the Grantee's incurred variable cost of providing the underground facilities and the estimated cost of constructing equivalent aerial facilities.

In this case, the subscriber pays a flat fee of $19.50 for normal connections and the cable operator's actual cost for underground service. This approach has merit. The variation among subscribers for overhead drops is small enough to make the ease of administering a flat rate system the dominant consideration, as with telephone installations. But underground construction can vary enormously, depending on the necessity of tunnels under sidewalks or streets, the extent of disruption to lawns and shrubbery, the nature of the soil, and other factors.

Charging an amount equal to actual cost may be a good practical solution—all the more so since the subscriber is given the option (in 1.2a above) to do his own trenching or to have some other contractor do it, in which case the charge by the cable operator is to be the same as for overhead service. These options may protect the subscriber against possibly excessive charges by the cable operator for trenching operations.

At the same time, an "actual cost" standard is not free of difficulty. "Actual cost" is an elusive measure, subject to disagreements between the subscriber and the cable operator and hard to define because of complexities in allocating the system's expenses among various functions. Moreover, subsection 1.2(a)'s provision for trenching by the subscriber may furnish only a limited alternative, because (a) subscribers may not know of the option or of a competent outside contractor, and (b) the cable operator might use petty defects in a job as an excuse for not laying cable in the subscriber's trench.

In view of these problems the franchising authority might consider an approach similar to the one suggested in relation to wiring apartment houses: setting a flat per-foot charge and allowing the cable operator to petition the appropriate tribunal for an increase.

The above franchise language also provides for an extra fee, equal to "actual direct cost," for extending a dropline more than 150 feet from the feeder. Any such limit is arbitrary, of course. But whether the number is set at 100, 150, or 200 feet, clearly some such provision should be included in the franchise so that subscribers who are extraordinarily expensive to serve bear their additional cost.

The Beverly Hills franchise is notable also in specifying that the installation charge for additional outlets on the same premises be lower than that for the first, to reflect the cost savings that accrue from installing multiple outlets from a common dropline into the home:

(b) Each Additional Outlet

   1.1 The charge shall be $5.00 per additional outlet if the subscriber's order for same is made not later than the time the first outlet is installed or reconnected or any other outlet is relocated.

   1.2 When the subscriber's order is made subsequent to the time specified in b. 1.1 above, the charge shall be $10.00 for the additional outlet and $5.00 for each further additional outlet, if any, ordered installed concurrently.

This stands in contrast to the provisions in the New York City franchises, which specify a flat maximum fee for *each* outlet:

Rates for Basic Service to Residential Subscribers shall not exceed the following amounts. . . For installation of each outlet, $9.95. . . .

In New York, then, the subscriber may pay $9.95 for each outlet he orders, whether one or five. However, since this figure is specified as a maximum, the cable operator could lower his rates for both the first and additional outlets, as long as he complies with the further provision that "All rates, charges, and terms or conditions relating thereto shall be non-discriminatory." In fact, cable operators in New York City follow the practice, widespread in the industry, of sometimes offering low "promotional" installation fees. This practice is permitted in New York City as long as during the period of time of the promotion the same low fees are offered on a non-discriminatory basis to all potential new subscribers.

MONTHLY SERVICE RATES, DISCONNECTION AND RECONNECTION CHARGES

In establishing a structure of monthly subscriber rates, it is important to distinguish between single and multiple dwellings and between TV service and FM radio service, as well as to specify charges for disconnection and reconnection. On a per-unit cost basis, serving a four- or five-unit dwelling is generally cheaper than serving a single-unit dwelling. These differences should be reflected in rates. Moreover, the fact that serving a second TV outlet or an FM outlet is cheaper than serving the first also should be reflected in rates, as noted above. An example of an interesting approach to a rate structure is drawn from the Beverly Hills franchise:

(a) Single family and duplex residential for TV and/or FM service:

First outlet	$5.00 per month
Additional outlets—each	1.00 per month
Disconnection of existing service	No charge

(b) Multiple apartments, hotels, motels, and non-residence for TV and/or FM service:

1.1	Number of outlets		Rate per month per outlet
	a.	1	$5.00
	b.	Next 9	3.75
	c.	Next 10	2.50
	c.	All over 20	2.00
1.2	Disconnection of existing service		No charge
1.3	The above rates shall apply only where all outlets are on the same premises, under one ownership and with billing to one customer.		

These provisions of the Beverly Hills franchise are interesting on several counts:

• The franchise specifies a monthly charge for each additional outlet substantially below the first, to reflect the difference in cost. (On this point the New York City franchises are in agreement; they also recognize the cost differences by specifying a lower monthly figure for each additional outlet ($1) than that for the first ($5)).

• In regard to the outlet charge, the franchise puts TV and FM service on the same footing, to reflect the fact that the cost of supplying them is the same; the cost of supplying only FM presumably is no different from the cost of supplying only TV, since both are carried on the same cable, which must be brought into the home and into an outlet. Although installation costs for FM and TV outlets are identical, a franchising authority might set a lower rate for an FM outlet, on the theory that an FM subscriber should not bear the costs of TV carriage and TV program origination on cable for which he does not receive the benefits.

• The franchise contains a sliding scale within which the monthly rate falls as the number of dwelling units on a single premise rises—again a reflection of the cost savings in serving points with high population density. Such a provision is especially important in encouraging the use of cable in low-income areas where multiple dwelling units prevail. This sliding scale applies, however, only when the premises are "under one ownership and with billing to one customer." Subcribers who live in condominiums or who are billed separately presumably would pay the full $5 per month, despite the fact that serving them would be substantially cheaper than serving single or duplex residences. In future franchises it would seem sensible to add additional rate categories. First, condominium owners should receive a sliding scale monthly charge below $5 but sufficiently above the monthly charges for "single billing to one customer" to offset the additional cost of billing each subscriber separately. Second, condominium owners or other residents who establish their own central billing system should receive the same sliding scale as multiple dwelling residents. Finally, the franchising authority might consider a separate set of rates for buildings which can be wired inexpensively—e.g., modern apartment houses with built-in cable or specially prepared ducts.

In contrast to the terms of the Beverly Hills franchise, the New York City franchises make no explicit distinction between single and multiple dwellings or between multiple dwellings of various sizes with respect to either the connection charge or the monthly service fee. Nor do they take into account length of connection line or underground versus aboveground construction. Such a flat-rate approach may be justified on grounds that the Borough of Manhattan is more homogeneous than is typical elsewhere: Virtually all dwellings are multiple units; once the feeder line is installed along the block, each dropline is relatively short; and practically all construction is underground. Moreover, although not required under terms of their franchises, the New York City cable operators on their own volition have offered quantity discounts (as filed with the Bureau of Franchises) to large multiple dwelling units where central billing is available.

• The Beverly Hills franchise requires no charge for disconnection of service. This also is a reasonable recognition of the practical problems in collecting disconnection payments, since the cable operator can do little to enforce payment. In any event, the small amount of money involved hardly would justify the operator's going to a small claims court or turning the bill over to a collection agency. The situation is all the more complex if the former subscriber moves out of town.[26] In the absence of a disconnection charge, however, the franchise should include a reconnection fee —which can be enforced more easily than a disconnection charge. If *neither* a disconnection or reconnection charge is imposed, the cable operator may incur very high costs in maintaining subscriber hookups, particularly in areas with high population turnover.

A striking example of the disconnection problem is that faced by the Kern Cable Company, which serves the unincorporated areas of Bakersfield, California pursuant to a franchise issued by Kern County. The following is taken from a letter from the cable operator to the Kern County Board of Supervisors:

> Kern Cable's records indicate that 25% of its customers moved during the year 1969. This is nearly double the turnover experienced by most other CATV systems.
>
> . . . .
>
> Kern is forced to set aside a minimum of one week a month during which as many as 16 servicemen do nothing but disconnect subscribers who have either moved or refused to pay their bills. A man must go to the door of the home and attempt to make a collection. If unsuccessful, he must then climb a utility pole and physically disconnect the drop line leading from the pole to the house. The entire process, including driving time, averages 20 to 30 minutes.
>
> Ironically, a disconnected subscriber can call a half-hour later and demand to be reconnected. He pays no reconnection charge; he has lost no deposit; all he has to do is pay his bill. Our installer must now go back out, climb the pole, reconnect the cable and go into the home to verify that service to the set has been restored. *Kern's present franchise makes no provision for recovery of the costs involved in either the disconnect or reconnect procedure.* [Emphasis added.][27]

The Beverly Hills franchise copes with the disconnection problem by specifying a *reconnection* charge:

(a) The charge shall be $5.00 per subscriber.
(b) The charge shall apply where the Grantee had previously served the premises, and facilities are substantially in place, but service was cancelled, suspended, or terminated for any good cause.
(c) The charge shall not be made in addition to the installation charge set forth in I. (a) 1.1 a, or II. above.

---

[26] One possible solution would be to require that the subscriber make an initial deposit in an interest-bearing escrow account, from which the cable operator could draw to satisfy any debts for disconnection of service.

[27] Letter, Kern Cable Company to Kern County Board of Supervisors, February 18, 1970.

(d)   The Grantee shall not remove, rearrange or otherwise disturb any of its existing facilities for the purpose of making a greater charge than would otherwise be applicable.

Thus, the Beverly Hills franchise does not permit a reconnection charge where the operator cancelled service without "good cause." This approach seems fair since it does not penalize the subscriber for the operator's mistake or misfeasance. At the same time, it could generate disputes in determining whether the operator had "good cause."

RELOCATION CHARGES

In addition to the preceding provisions, the franchise should specify charges for relocating outlets within the same premises. The Beverly Hills franchise is notable in providing that relocation of a second and additional outlet is to be priced lower ($5) if related to other services or performed at the same time:

IV.   RELOCATION CHARGE

(a)   The charge shall be $7.50 per relocated outlet, except that where the subscriber's order for same is made not later than the time an additional outlet is installed, the subscriber's service is reconnected, or the first outlet installation is made, the charge shall be $5.00 per relocated outlet.
(b)   In the event that more than one outlet is ordered relocated concurrently, the charge for the first such outlet shall be as set forth in IV. (a) above, and the charge for each additional relocated outlet shall be $5.00.

This also stands in contrast to the New York franchises, which specify as a maximum ". . . for moving and reconnecting an outlet, $9.95," regardless of the number of relocations and regardless of the fact that relocating an outlet is less costly than installing a new one. Again, the cable operator on his own volition may make nondiscriminatory downward adjustments in his rates to reflect these differences in cost.

BILLING AND PAYMENT PROCEDURES

Franchises typically specify procedures by which payment is to be made, including payment for a portion of a month's service. As one example, the Beverly Hills franchise specifies:

Prorating for Less than One Month's Service

The monthly rate shall be prorated on the basis of the number of days in the period for which service was rendered to the number of days in the billing period.

Billing and Payment

1.1   The bill for the monthly rate may be rendered in advance. Such bill is due and payable not more than five (5) days in advance of the period during which service is to be furnished.
1.2   The billing period shall not exceed two (2) calendar months.

Though billing a customer for service in advance obviously guarantees payment and is practiced by many utilities, it vests more power in an already quasi-monopolistic entity. Moreover, a system of advance payments makes rebates for poor service all the more difficult; it transforms them from rate reductions to refunds, and problems may arise in assuring prompt issue of refund checks.

PROVISION FOR TEMPORARILY REDUCED CHARGES

Occasionally, a cable operator will offer special reduced rates for promotional purposes. During a subscriber sign-up drive, for example, he may waive the cable connection fee temporarily. In some cases he even will offer a free 30-day trial hookup in the area of the promotional drive. Some cities, such as Beverly Hills, specifically provide for such possibilities.

> At the option of the Grantee, any of the foregoing charges may be reduced by any amount, but only under the following conditions:
>
> a. The subscriber makes a bona fide application for service not later than 30 days after the Grantee's local feeder cable is initially placed in service and such cable is reasonably accessible to the subscriber's premises; or
> b. The Grantee applies· for and receives from the City Council specific time limited authority to make reduced charges.
> c. Reduced charges, if offered, shall be on a non-discriminatory, non-preferential basis.

The franchise thus strikes a reasonable balance by allowing the cable operator to run a promotional campaign while preventing him from using an alleged promotional campaign as a cover for discriminatory rates.

REDUCED RATES FOR SPECIAL CLASSES OF USERS

The requirement of "nondiscriminatory" rates appears in most franchises. There has been much debate, however, about whether the availability of cable should be promoted among low-income groups by means of special preferential rates. A leading problem is defining in a clear and workable fashion who does and does not qualify. A threshold assumption must be that no formula ever can be truly fair. As the courts have found through long and bitter experience in administering the equal protection clause of the federal Constitution, any draftsmanship inevitably will be overinclusive in some respects and underinclusive in others.[28] Thus a preferential rate for welfare recipients inevitably will include some people who can afford to pay and exclude other people who cannot. This simple fact of life does not rule out preferential rates; it must be accepted, however, before any realistic analysis can begin.

If welfare recipients do receive special rates, the constant turnover in the welfare rolls may create some difficulties. Whenever a person went on or off the rolls, the welfare agency would have to notify the cable operator—who in turn would need to adjust his billing. On the one hand, this might generate substantial costs for the cable operator and resentment among subscribers. On the other, welfare eligibility has been used as a criterion for other services in the past, and may not create substantial transactional costs.

---

[28] Tussman and Tenbroek, *The Equal Protection of the Laws,* 37 *Calif. L. Rev.* 341 (1949).

Another approach, included in the Pomeroy, Ohio franchise, is to give a special rate to the head of any household who is either more than 65 years old or permanently disabled. In contrast to the regular monthly fee of $5.50, the Pomeroy franchise provides:

"Senior Citizens Service": Homes in which the head of the household is 65 years old or older ("Head of household" being determined by definitions of the U.S. Bureau of the Census and U.S. Internal Revenue Service) — $3.50 per month.

"Disability Special Service": Homes in which the head of the household, as defined above, is certified as permanently and totally disabled, under definitions of the U.S. Department of Health, Education and Welfare, or is suffering from a service-connected disability exceeding 60 percent, as defined by the U.S. Veterans Administration—$4.00 per month.

One danger of this approach is that if sufficiently large numbers of subscribers qualify for the reduced rate, the cable operator's receipts may not cover his total costs. In recognition of this danger, the Pomeroy franchise includes an additional provision:

It is further understood that in the event the total families in the "Senior Citizens Service" and "Disability Special Service" categories shall exceed 40 percent of the total of paying subscribers, the Company shall have the right to increase rates for these categories, by filing an amended schedule of such rates with Council.

Another question with preferential rates is the extent to which monthly fees within a franchise area should be adjusted to reflect underlying cost differences that result from variations in population density. All else being equal, the cost of cabling a densely populated area is less per dwelling unit than that of cabling a sparsely populated area. As a social issue the problem is all the more pressing, since at least in cities low-income groups tend to live in relatively densely populated areas. A single uniform rate creates the danger that the densely populated low-income central portion of the city would subsidize the higher-income fringes. The bulk rates for multiple dwelling units, as in the Beverly Hills franchise noted above, help to alleviate this problem.

Most franchises specify a single rate within the entire franchise area for any given class of users, such as single-family or two-family residences. If the franchise area is large, encompasses a wide variety of housing densities, and hence has large variations in underlying costs, the franchising authority should consider the feasibility and desirability of a differential rate structure. As in the preferential rate discussion above, however, great care would be required to establish definitions of classes of users that are unambiguous and do not generate resentment and misunderstanding. One possibility would be to establish a preferential rate for people living in census tracts with more than a specified number of people per street-mile. For a medium-size city, 150 to 170 people per street-mile is typical in the most densely populated portions, and 100 or so in the periphery. In such a case the franchise authority might consider lowering rates at a break-point of 130 to 140 persons per street-mile within census tracts, as defined by most recent census data.

This approach demands careful scrutiny of the area's actual composition, however, since density alone does not determine overall cost per subscriber. Among considerations bearing upon overall costs of wiring and providing service are the placement and age of buildings; the extent of turnover among subscribers due to differences in mobility among groups; and the extent to which collection costs and the level of market penetration, including the sign-up for special services, varies among areas as a consequence of differences in family income and other factors.

ESTABLISHING AND ADJUSTING THE RATE STRUCTURE

A critical task facing franchising authorities is the determination of a rate structure and of mechanisms for modifying it. Until recently, the rate for "basic service"—standard one-way television service over cable—has fallen within narrow bounds. Virtually all franchises have specified a rate in the range of $4 to $6 per month.[29] Likewise, the rate for set-top converters to increase the number of cable channels also has fallen in a narrow range, with a typical rate of $1 per month.

But as new services are added, the problem of rate setting and adjusting will become increasingly complex. If movies, education, data information storage and retrieval, alarm services and other applications occupy special channels, they will create problems not only as to their pricing, but also of assessing their impact on revenue requirements for the basic and other services. For example, if the cable operator collects substantial revenues from a special pay-movie channel, he may be able to reduce the rate for basic service substantially below the traditional $4 to $6 per month. Indeed, if a host of new services become lucrative it eventually may become possible to reduce the basic monthly service rate to zero—in effect, giving the cable subscriber free conventional one-way television service and using revenues from special add-on services to cover the total cost.

SETTING RATES AND CONNECTION CHARGES FOR BASIC AND ANCILLARY SERVICES

Establishment of equitable initial rates and fees depends on the conduct of the franchising process itself, as described in Sec. II of this report. Once the franchising authority decides which applicant is to be awarded the franchise, it simply might incorporate its proposed schedule into the final franchise document—assuming that it has opened rates up to bidding in the first place. Though the temptation is great to negotiate further rate adjustments after the franchise award, the franchising authority should refrain from doing so; this type of ex post negotiation would deny other applicants a fair chance to make further bids. This difficulty of negotiating with the successful bidder is one of the areas of inflexibility accompanying the "competitive bid and award" approach, as distinguished from the "negotiation" approach dismissed in Sec. II of this report.

REVIEW AND REVISION OF RATES

In addition to setting initial rates, the franchising authority must devote serious attention to periodic rate modifications. On the one hand, general inflationary

---

[29] A random sample of 25 systems listed in the *Television Factbook* (1971-72 edition) discloses that two have a monthly subscriber charge of less than $4 and that one has a charge exceeding $6.

pressures may require periodic rate increases if the cable operator is to maintain a viable business. On the other, continued technological advance, new economies in operating procedures, and development as well as expansion of programming markets may permit rate reductions. Given these possibilities, the franchising authority should adopt language that will make necessary rate adjustments relatively easy. The language should be specific enough to avoid misunderstandings and the uncertainty of long and frustrating hearings, rehearings, and reviews. At the same time, the franchise should not simply ignore the issue by omitting all language with regard to mechanisms for change. Unfortunately, many franchises are deficient in these respects.

One example is the Los Angeles franchise, which leaves open the criteria by which the franchising authority is to regulate rates:

REGULATION OF RATES AND SERVICE. The Board shall have the power to regulate rates and service to the extent and in the manner as provided in the City Charter and ordinances, and the Grantee by its acceptance of this franchise agrees to comply with every such order and regulation.

The nature of "every such order and regulation" remains to be seen. If little disagreement arises between the city and the cable operator, the absence of specific criteria for judging rates will pose no serious problem. Yet one easily can imagine circumstances under which the cable operator's agreement to "comply with every such order and regulation" could give rise to serious problems.

As another example, the Akron franchise states that:

When this Franchise takes effect, the rates herein contained shall be subject to revision by Council after a period of ten (10) years elapses as provided in Section 46 of the Akron City Charter.

The Company from time to time may make changes in or additions to the rate schedule, provided, however, that the change or additions shall become effective sixty (60) days after the filing of notice of such change with the City, unless disapproved within such period for good cause shown by the Council of the City.

In this case it is not clear what criteria the Council would use in revising rates after the ten-year period; and while the cable operator is free to change the rate schedule after filing due notice, the basis for approving or disapproving the changes remains unclear.

Moreover, "approval by acquiescence" may be dangerous, since under some states' law the franchising authority's failure to take action may not be subject to judicial review. Accordingly, the franchise should provide that any rate change accepted by the franchising authority is a final order—and thus reviewable by the courts.

Another example is the language in the Columbus, Ohio City Code:

No action shall be taken by the Council with respect to rate reduction and/or modifications in rate structures unless the operator has been given a written notice at least ninety (90) days in advance of the effective date contemplated by Council *and not until the operator has been given every opportunity to be heard by Council will final action be taken* (Emphasis added).

Here serious problems can arise in defining "every opportunity." One easily can imagine months of debate, delay, and frustration in case of serious disagreement, while the cable operator and the city attempt to explore "every opportunity" to reach some kind of agreement.

Other franchises take what might be called an "asymmetrical approach" to the issue of review and revision, by in effect making rate reductions easier than rate increases. The New York City franchises are a leading example:

> The Board may reduce rates for Basic Service at any time after five (5) years from the effective date of this contract and rates for Additional Service after eight (8) years from the effective date of this contract upon a determination, made after a public hearing following notice to the Company, that such rates or a particular rate can be reduced without impairing the ability of the Company to render service and derive a reasonable profit therefrom.

This provision is sensible in cases where the cable operator is able to enjoy operating economies, take advantage of cost-cutting new technologies, and develop new services to provide additional revenues and a broader base for his overhead costs. In early years he can retain the full benefit of whatever cost savings and additional revenues he is able to achieve; and this would have the beneficial effect of giving him an incentive to operate as efficiently as possible.

But the New York City franchise has no comparable provision for cases where rate *increases* might be necessary. If faced with clear and obvious financial difficulty, the cable operator presumably could petition the franchising authority for a rate increase under the provision quoted below; but the absence of specific language delineating the criteria for rate increases could be a source of difficulty:

> The Board may at any time increase or decrease any rate, require discontinuance of any scheduled service, or revise or delete any term or condition applicable thereto upon a determination, made after a public hearing following notice to the Company, that a particular rate, service or term or condition (1) explicitly or implicitly violates this contract or (2) has the effect of unreasonably restricting the use of Public Channels.

As noted previously, the franchising authority can choose among many rate-control strategies; but whatever standard it ultimately adopts should be as clear as possible in the franchise, for the benefit of both the cable operator and the public.

The process of public hearings is an important input in achieving this goal. Giving effective notice to the public is considerably easier at the rate-modification stage than at the initial franchising stage, since the cable system usually will be in operation. The franchise authority might rely heavily on the system's own facilities to give notice, by requiring the operator to cablecast frequent announcements of any pending rate hearing. In addition, the franchising authority might require at least some other notice, to cover citizens who do not subscribe to the cable system but who are concerned with its rates.

Public hearings on rate changes should be conducted like those in the initial franchising process. Obviously, the franchising authority must provide time not only for citizens to be heard, but also for the cable operator to present his position and rebut others' presentations. In this proceeding, the city attorney or other local legal officer may take an affirmative, quasi-prosecutorial role—much as the Broadcast Bureau of the FCC sometimes does in license renewal hearings.

RATES FOR NEW SERVICES

The final question is pricing wholly new services as they arise. How and to what extent should the franchising authority—and can the franchising authority, under the FCC's rules—control the rates for special pay movies, alarm services, information storage or retrieval, and other services? In some of these cases, such as pay movies, the strong competitive pressure from theaters and commercial television may obviate the need for explicit public control. Other services may be highly monopolistic, however, and also vital to the needs of the changing American society. For now, a relatively simple statement in the franchise may be sufficient to leave the door open for necessary ratemaking. The Beverly Hills franchise provides a good example:

Any service not covered by scheduled charges and deemed desirable may be offered and a rate may be initially established by the Grantee. Such rates may be reviewed and thereafter established by the City Council.

The FCC may provide some future guidance in setting rates for new services. The Commission has indicated that it plans:

... at a later date to institute a proceeding with a view to assuring that our requirement of capacity expansion is not frustrated through rate manipulation or by other means. This proceeding will also deal with such open questions as rates charged for leased channel operations.[30]

ALLOCATION OF CHANNELS

The FCC has been concerned greatly about pressure on cable operators to offer large numbers of free channels for educational and other public uses. Although the cost of one channel may be small, it is not zero. The more free channels offered, the greater is the cable operator's cost burden—which usually will be passed on to subscribers in terms of less service or higher fees. At the same time, the FCC has concluded that *some* free channel capacity should be made available for development purposes and for assured public access. Hence it has specified that, in addition to leased channels, three channels—and only three channels—be set aside free of charge for special purposes:

... cable television systems will have to provide one dedicated, noncommercial public access channel available without charge at all times on a first-come, first-served nondiscriminatory basis and, without charge during a developmental period, one channel for educational use and another channel for local government use.[31]

The grandfathered New York City franchises provide an example of provisions that generally would be invalid under the new FCC policy. Of the initial 17 channels, two were set aside for the city and two for public access. Moreover, according to these franchises:

As the channel capacity of the System is increased beyond seventeen (17) channels new channels shall be allocated in the following sequence: one (1) City Channel, two (2) Public Channels and three (3) Additional Channels. The Director of Communications shall designate dial locations as he deems appropriate for all new channels.

[30] *Cable Television Report and Order* at 3276. The FCC's present rules require a cable system to publish a rate schedule for its leased channels, however, which will make covert rate discrimination more difficult. 47 C.F.R. § 76.251(a)(11)(iii) (1972), printed in *Cable Television Report and Order* at 3289.
[31] Ibid. at 3269.

Under the FCC rules a new franchise presumably could require free of charge only one city channel, one public access channel, and one educational channel.

The only exception the FCC recognizes is a possible waiver if a community and cable operator jointly make a special showing relating to experimental uses of the cable system:

Because of the federal concern, local entities will not be permitted, absent a special showing, to require that channels be assigned for purposes other than those specified above. We stress again that we are entering into an experimental or developmental period. Thus, where the cable operator and franchising authority wish to experiment by providing additional channel capacity for such purposes as public, educational, and government access—on a free basis or at *reduced charges*—we will entertain petitions and consider the appropriateness of authorizing such experiments, to gain further insight and to guide further courses of action. [Emphasis supplied.][32]

Any special showing presumably would be made by way of a petition for special relief, as previously discussed. The necessary facts and documentation still are not clear, since the Commission has not yet passed on any such petitions. The petition presumably should show that the franchise area has a tangible need for additional channels, however, and should be documented by affidavits testifying to either an inability to get time or to a desire to use time. Another study in this series explores the requirements for waivers in more depth.[33]

Moreover, these provisions leave open several other questions. First, to what extent may the local franchising authority require the cable operator to make available channels on a preferential basis—i.e., for a charge that covers at least the incremental or out-of-pocket cost of supplying the channels but not a proportionate share of the system's overhead. For example, may the franchising authority authorize additional educational channels at rates high enough not to impose an additional cost burden on other users, but below those for pay movies and sports? A pricing practice in which some services bear much more overhead than others is common to virtually all industries, including the telephone industry. As a case in point, nighttime telephone calls are priced at rates lower than daytime calls, yet at rates sufficient to cover the additional costs of using otherwise largely idle facilities. To the extent that nighttime telephone revenues cover *some* portion of overhead, daytime users actually benefit from nighttime service, even at lower rates. According to the language in the *Cable Television Report and Order,* however, these special reduced rates apparently would be authorized only on the basis of a special showing of experimental, rather than routine use. This conclusion is strengthened by the FCC's more recent pronouncement on reconsideration of the rules:

The NAEB [National Association of Educational Broadcasters] urges us to amend our rules to enable leased channels to be used for educational purposes at lower rates and to provide that, at the termination of the free five year developmental period for educational access channels, rates be kept at a minimum. As stated, we are entering into a period of experiment. The access rules will, without question, require further study and future deliberations. The question of access channel rates is but one of the matters which we will have to confront again. Our initial feeling in this matter is to avoid any form of preferential policy with regard to who may use and what must be paid for access channels. For the present, we deem it desirable to allow the experiment to proceed apace.[34]

---

[32] Ibid. at 3271.
[33] See Chapter 5.
[34] *Reconsideration Opinion and Order,* at 13859.

Second, if the city does not have additional free channels, is it entitled to other forms of preferential treatment, such as in service rates? For example, the New York franchises specify that:

> The charge for Basic Service [relay of broadcast signals, origination programming, and access channels] to any board, bureau or department of the City, or other governmental body, or any public benefit corporation for Basic Service in addition to any such service provided free of charge shall be seventy-five percent (75%) of the charge for Basic Service to Residential Subscribers.

> The Company shall provide the City with Additional Service [new services] at a charge to the City not to exceed sixty-six and two-thirds percent (66-2/3%) of the tariff charge for similar service offered by the appropriate communications common carrier operating within the District.

Nothing in the rules directly seems to prohibit preferential treatment in terms of *monthly subscriber rates.* On the one hand, the FCC might invalidate these types of preferences also, on the theory that they are inconsistent with its general position against preferences. On the other hand, the FCC's previously discussed broad grant of rate-control power to local governments might allow preferences in subscription rates to both government and nongovernment users.

Third, may the franchisor require the cable operator to wire public facilities free of charge or under preferential conditions? In the absence of FCC language to the contrary, franchising authorities presumably will continue to have leeway in requiring the operator to wire specified facilities. One example is an extensive list drawn from the Beverly Hills franchise:

> The undersigned agrees to provide complete CATV service anywhere within the city limits to any public facility now owned or acquired in the future by the City of Beverly Hills or now owned or acquired in the future by the Beverly Hills Unified School District subject to the following conditions:

> (a) All underground conduit or conduit within any public building will be installed without expense to the undersigned.
> (b) The cost of material and labor for cable installation in excess of 1,500 feet in length, measured from the closest point of each service, will be installed without expense to the undersigned.
> (c) No material or labor charge, installation charge, monthly charge or other charge (except as otherwise provided in Paragraphs (a) and (b) of this paragraph) shall be made to the city or the school district for such CATV service.
> (d) Right of way will be provided at no expense to the Grantee.
> (e) Authority for requirement of such service shall be by resolution of the Council.
> (f) The locations of facilities now owned by the City which may be designated for service under requirements of this section are shown on Map Number 2.
> (g) The locations of facilities now owned by the Beverly Hills Unified School District which may be designated for service under requirements of this section are shown on Map Number 3.

This type of provision could constitute a substantial cost burden on the cable operator. In view of the FCC's concern about such burdens, the Commission should determine whether this is a loophole that should be plugged. The Commission appears willing, however, to allow free services for a limited number of municipal functions such as fire and police agencies.

Fourth, an ambiguity arises with respect to the cable operator's obligations to supply three channels where one headend serves more than one community. The FCC requires that "*cable television systems* will have to provide one . . . public access channel, one channel for educational use and another channel for local government use."[35] But in many situations communities will find it

---

[35] *Cable Television Report and Order* at 3269 (emphasis supplied).

mutually advantageous to franchise a cable operator jointly in order to exploit economies of scale.[36] Must the cable operator then supply three separate channels for *each* of the communities he serves from a single headend, or could the communities share the use of each channel? The FCC has noted that:

> ... we are not unmindful of the existence of multiple systems served by a single headend. In most of these situations, each system has the same channel capacity, and carries the same broadcast programming. The "system" as a whole is not designed to carry program material selectively to each component system. The ability of an existing conglomerate of systems to comply with the access channel requirements will necessarily vary with the proximity of the component systems, the basic design of the system, and, of course, the channel capacity. Clearly, we cannot establish a rule of general applicability in this area. To the extent possible, however, within the technical and geographic parameters of the systems involved we intend to safeguard the integrity of our access requirements. This can best be done if, during the certificating process we are provided with *sufficient detailed information concerning the systems' ability to comply*. Again, we will require compliance to the greatest possible extent. In some cases it may be possible for individual systems to share channel time. If this is the case we may be persuaded for instance that at least 2 shared public access channels will suffice for some conglomerate systems. Where boards of education are under the same jurisdiction, the problems may be alleviated. Local governments may agree to share time on one or two channels. We must, however, *be given as much information in these respects, as possible, together with specific proposals on the part of the systems* [Emphasis added]. Until we receive such material certificates will not be issued.[37]

At present, the FCC seems willing to let cable systems share access channels among neighboring communities served by a common headend.

Supplying each community with an educational, governmental, and public access channel could constitute a substantial burden on a cable operator who serves three, four, or five communities. Moreover, small communities—in the range of 5,000 to 10,000—may not use these channels very effectively. At the other extreme, where a single system serves a city of a million people, the three channels would be grossly inadequate. In short, communities that contemplate sharing headends must submit a specific proposal to the FCC as indicated above.

PAY PROGRAMMING

In response to pressure from local broadcasters and motion picture exhibitors, many franchising authorities have taken a hostile position toward pay programming on cable systems. For example, the New York City franchises provide that "[T]he Company shall not engage in Pay Television, nor shall it deliver signals of any person engaged in Pay Television, unless and until affirmatively authorized by the F.C.C."

Although opposition has been expressed within the FCC to any form of pay programming—whether broadcast or cablecast—its recent rules specifically authorize pay programming, subject to some restrictions on program sources.[38] This codifies an earlier ruling in which the Commission responded to the situation in New York City and "affirmatively authorized" pay programming.

[36] For example, one recent study has shown that a number of suburbs ranging in population size from 3,000 to 50,000 each would find it in their interest to share the headend and studio facilities in order to reduce per-subscriber cost. Johnson et al., *Cable Communications in the Dayton Miami Valley*, Paper 1, pp. 9-13.

[37] *Reconsideration Opinion and Order* at 13860.

[38] FCC, *Notice of Proposed Rulemaking*, 37 Fed. Reg. 3192 (1972).

FACILITIES FOR PUBLIC ACCESS TO SYSTEM

Since 1969 the FCC has required systems with more than 3,500 subscribers to originate programming "to a significant extent as a local outlet."[39] In addition, the FCC's new rules require one channel for public access, one for government, and one for education. With respect to the latter two, the cable operator's obligation clearly extends only to supplying a channel, as opposed to studio equipment or other facilities on the premises of educational and government institutions. It is not clear how much leeway franchising authorities will have, however, in requiring the operator to wire specific public facilities free of charge, as noted below.

The public access channel poses a different problem, since it involves studio facilities that are under the control of the cable operator. As the FCC noted:

> We believe there is increasing need for channels for community expression, and the steps we are taking are designed to serve that need. The public access channel will offer a practical opportunity to participate in community dialogue through a mass medium. A system operator will be obliged to provide only use of the channel without charge, but production cost (aside from live studio presentations not exceeding 5 minutes in length) may be charged to users.[40]

The franchise should state the general specifications for program origination equipment, such as:

- Should the equipment be color or monochrome?
- What quality of equipment should be used—e.g., half-inch tape, one-inch tape, etc.?
- Should the equipment be compatible? If so, what standard should be adopted?
- How many cameras, recorders, and playback units should be required?
- What type of technical assistance should be required?

Obviously enough, no written document can specify how to run a television studio, especially in light of a constantly changing technology. Nevertheless, a fairly detailed set of initial provisions will minimize later conflicts; a reservation of power clause will enable the franchising authority to impose additional and more specific requirements as necessary.

In doing so, one would expect that the larger the subscriber base, the more elaborate the facilities to be specified. Here as elsewhere, franchise officials should not saddle the cable operator with costly obligations in the absence of clear public benefits.[41]

Requiring production equipment may violate the FCC's ban on a franchise fee of more than three percent, absent a waiver. The FCC might consider the cost of the

---

[39] 47 C.F.R. § 76.201 (1972), printed in *Cable Television Report and Order* at 3287. The requirement first was imposed by the *First Report and Order*, 20 F.C.C. 2d 201 (1969) and was reaffirmed in the *Memorandum Opinion and Order*, 35 Fed. Reg. 10901 (1970). When the new rules were promulgated, the existing origination requirement was taken over virtually verbatim.

[40] *Cable Television Report and Order* at 3270.

[41] Alternative equipment lists required for particular levels of local program origination are contained in several previous studies, including Charles Tate (ed.), *Cable Television in the Cities: Community Control, Public Access, and Minority Ownership*, The Urban Institute, Washington, D.C., 1971, pp. 42-51.

production equipment and of the studio maintenance as part of the consideration for the franchise and thus add it to any other franchise fee expressly exacted. As a result, the franchising authority may need to petition the FCC for a waiver. To be successful, the franchising authority presumably would need to show that the production equipment is related to a bona fide regulatory or experimental program, as will be discussed later.

MINIMUM CHANNEL CAPACITY

Most franchises specify the capacity of the system, and some franchises specifically provide for expanding capacity in accordance with the demand for channels. One example is drawn from the model franchise of the MVCOG:

The grantee has in its proposal agreed that the cable television system shall have forty (40) channels initially. The grantee shall provide additional channel capacity in accordance with F. C. C. regulations. Failure to comply with this provision shall be a material breach of this franchise and shall be a sufficient cause for revocation of this franchise.

The FCC states that:

We wish to proceed conservatively, however, to avoid imposing unreasonable economic burdens on cable operators. Accordingly, we will not require minimum channel capacity in any except the top 100 markets. In these markets, we believe that 20 channel capacity (actual or potential) is the minimum consistent with the public interest. We also require that for each broadcast signal carried, cable systems in these markets provide an additional channel 6 MHz in width suitable for transmission of class II or class III signals.[42]

Thus, the design of the system, including trunk lines, feeders, droplines, and amplifiers must accommodate a minimum 20-channel capacity, though the actual number of operating channels will depend on the number of broadcast signals, leased uses, and other factors. If the cable system carries more than 10 broadcast channels, its capacity must be increased beyond 20 to provide an equal number of nonbroadcast channels.[43]

Thus franchising authorities in major markets clearly now may not allow a system to have less than 20 channels. A question arises as to whether a franchising authority may require *more* than 20 channels—e.g., 30 or 40. Although the FCC's rules explicitly prohibit regulation of "the number or manner of operation of *access* channels," [emphasis supplied][44] the rules relate only to these channels and not to the system's total channel capacity. As a result, a franchising authority probably is free to require a channel capacity greater than the FCC's minimum, if it can demonstrate any rational basis for the requirement. In reconsidering the rules, the FCC commented that:

[42] *Cable Television Report and Order* at 3269. For additional commentary see Rivkin, op. cit., Sec. 3.332.

[43] *Cable Television Report and Order* at 3270.

[44] 47 C.F.R. § 76.251(a)(11)(iv) (1972), printed in *Cable Television Report and Order* at 3289.

The question has arisen whether we have pre-empted the area of channel capacity so that local governmental entities could not require more than twenty channel capacity or more than required under the equal bandwidth rule, Section 76.251(a)(2). We believe that our requirement for expansion of channel capacity will insure that cable systems will be constructed with sufficient capacity. However, if a local governmental entity considers that greater channel capacity is needed than is required under the rules, we would not foreclose a system from meeting local requirements upon a demonstration of need for such channel capacity and the system's ability to provide it.[45]

INTERCONNECTION OF SYSTEMS

A question also arises about whether local franchising authorities can *require* interconnection of systems. In its *Cable Television Report and Order* the FCC is silent about this issue. But the previously discussed ban on local "rules concerning the number or *manner* of operation of access channels" (emphasis supplied) could be interpreted as restricting the power to require interconnection, since interconnection could affect the *manner* of operation. Interconnection is certainly important in supporting new services that need a large subscriber base. Though the FCC has not yet passed on the issue, its comments on reconsideration of the rules indicate that the issue is at least still open:

The National Association of Educational Broadcasters is concerned with how cable is to develop to assure the interconnection of franchise areas (regionally or statewide) and the adequate planning of equitable service expansion from urban to rural areas. Petitioner argues that local officials may not be able to meet such a challenge for compatible development and interconnection across political boundaries. Again, we feel that it would be premature to codify such rules as the petitioner suggests. However, we do agree with the NAEB that such guidelines should be identified as a priority problem for the Cable Television Advisory Committee on federal-state/local relationships.[46]

If local authorities ultimately are permitted to require interconnection, they should keep in mind a caveat: though very important for services requiring a large subscriber base, interconnection involves a cost that could get out of hand if the franchise language is written too broadly. An example of this risk can be seen in the language of the American Civil Liberties Union's Model Code:

Each franchisee *must* interconnect its cable system with all other contiguous cable systems, and may interconnect with any other system or service,

---

[45] *Reconsideration Opinion and Order* at 13858, n. 25. The FCC continues in the same footnote: "A similar question has been raised with respect to two-way capability. We find no reason why a cable operator wishing to experiment with a more sophisticated two-way capability than that which we have required should be precluded from doing so. However, we do not believe that franchising authorities should require more than we have provided for in our rule because it is possible that any such requirement will exceed the state of the art or place undue burdens on cable operators in this stage of cable development in the major markets. Where a franchising authority has a plan for actual use of a more sophisticated two-way capability and the cable operator can demonstrate its feasibility both practically and economically we will consider, in the certificating process, allowing such a requirement."

[46] Ibid. at 13862.

in such a way that subscribers can receive *all* channels of all interconnected systems at any time. [Emphasis supplied.]

The cost of interconnecting systems to provide *all* channels of all contiguous systems could be very great. For example, if each system has 20 channels, the interconnection network would require a 20-channel capacity in *each* direction between contiguous systems—an endeavor that could be extraordinarily expensive and also require perhaps more radio spectrum space (if microwave links are employed) than is available.[47]

As one example of past kinds of interconnection provisions, the New York City franchises specify:

> For the purpose of permitting the transmission of signals throughout the City the Company shall interconnect its System with any other broadband communications facility authorized by the Board to operate in an adjacent district. Such interconnection shall be made within sixty (60) days from the effective date of this contract with the System presently operated in the southern portion of Manhattan by Sterling Information Services Ltd. Within four (4) years the Company's System shall be capable of interconnection with any broadband communications facility authorized by the Board in an adjacent district and with any adjacent community antenna television system (as defined by the F.C.C.) outside the City; actual interconnection may be ordered by the Director of Franchises upon reasonable terms and conditions.

New York City's approach thus is somewhat more selective than the ACLU's, since it requires interconnection only with the City's other systems and does not specify carriage of an inflexible number of channels.

FRANCHISE FEES

Because of today's prevalent financial pressures, many cities view cable television as merely another convenient source of revenue. Competing franchise applicants often find themselves offering higher and higher fees to sweeten their bids.

Yet the FCC and many other groups have expressed great concern about franchise competitions' tendency to be based on fees. The operator can be squeezed so badly that his ability to serve the public is curtailed severely. The FCC notes that:

> ... we believe some provision is necessary to insure reasonableness in this respect. First, many local authorities appear to have exacted high franchise fees more for revenue-raising than for regulatory purposes. Most fees are about 5 or 6 percent, but some have been known to run as high as 36 percent.
> The ultimate effect of any revenue-raising fee is to levy an indirect and regressive tax on cable subscribers. Second, and of great importance to the Commission, high local franchise fees may burden cable television to the extent that it will be unable to carry out its part in our national communications policy. Finally, cable systems are subject to substantial obligations under our new rules and may soon be subject to congressionally-imposed copyright payments. We are seeking to strike a balance that permits the achievement of Federal goals and at the same time allows adequate revenues to defray the costs of local regulation.
> ... It is our judgment that maximum franchise fees should be between 3 and 5 percent of gross subscriber revenues. But we believe it more appro-

---

[47] For a discussion of interconnection costs, see Johnson et al., op. cit., Paper 1, pp. 29-31.

priate to specify this percentage range as a general standard, for specific local application. When the fee is in excess of 3 percent (including all forms of consideration, such as initial lump sum payments), the franchising authority is required to submit a showing that the specified fee is appropriate in light of the planned local regulatory program, and the franchisee must demonstrate that the fee will not interfere with its ability to meet the obligations imposed by our rules.[48]

The FCC places this limitation only on subscriber revenues. In its rules it says nothing about limitations on fees from leased channels, advertising, and other sources of revenue. More recently, however, the Chief of the Cable Television Bureau has stated in answer to a question from a cable operator that franchise fees are to be computed only on the basis of regular subscriber revenues:

Q. May a franchising authority impose a franchise fee based upon revenues derived from "auxiliary" services such as advertising revenues, leased channel revenues, pay cable revenues, etc.?

A. No—Subscriber revenues are considered to be those revenues derived from regular subscriber services—i.e., the carriage of broadcast signals and required non-broadcast services.[49]

Unless modified as circumstances warrant, this policy could create problems in the future. If new services thrive, it may be possible for cable operators to reduce regular subscriber rates, perhaps eventually to zero, with most or all revenues generated by special services. Unless franchise fee ceilings take into account the development of new services, there could be a forced reduction in fees through time. For the moment, franchising authorities will need to keep this restriction carefully in mind. The state of uncertainty they are left with will be clarified through time as the FCC issues, or refuses to issue, certificates of compliance.

An example of the problem posed by the new FCC rules is the language of the grandfathered New York City franchises:

As compensation for this franchise the Company shall pay the City amounts equal to the following:

(i) Five percent (5%) of its Gross Receipts from provision of Basic Service to Residential Subscribers, starting on the effective date of this contract, but not less than the following minimums for the calendar years specified:

1971	$75,000
1972	100,000
1973	125,000
1974	150,000
1975	175,000
1976	200,000
1977	250,000
1978	300,000
1979-1989	350,000

In 1990, $350,000 pro-rated from January 1 to the twentieth anniversary of this contract; and

(ii) Ten percent (10%) of all its other Gross Receipts; and

(iii) When and if Pay Television is authorized, the percentage of the Gross Receipts therefrom, or other compensation, determined pursuant to Section 4(1).

[48] *Cable Television Report and Order* at 3276-77.
[49] Letter No. 4000U from Chief, Cable Television Bureau, to Edward M. Allen, August 11, 1972.

When and if so authorized, the Company shall not engage in, nor deliver the signals of any person engaged in, Pay Television until the amount of compensation payable to the City by the Company or other person engaged in Pay Television has been fixed by the Board. Such compensation shall not exceed twenty-five (25) percent of the Gross Receipts attributable to such Pay Television.

If this approach were followed in drafting new franchises, it would raise several questions for local authorities:

- The 5 percent of gross receipts from basic service exceeds 3 percent and thus would require a special showing that the fee "is appropriate in light of the planned local regulatory program" and "that the fee will not interfere with its [the system's] ability to meet the obligations imposed by our rules." New York City's relatively elaborate cable regulatory framework—including a new Office of Telecommunications—may provide a sufficient basis for justifying the 5-percent fee.[50] It remains to be seen whether other cities, particularly smaller communities, will have regulatory programs justifying an additional fee.

- The inclusion of fees on gross receipts from new services (10 percent in New York City's case) would run afoul of the policy discussed above.

- Similarly, fees on pay television (a maximum of 25 percent in the case of New York City) would be contrary to FCC policy.

- To follow New York City's approach in specifying flat minimum fees for the duration of the franchise conceivably might run afoul of the FCC rules, under which the franchise fee includes "all forms of consideration, such as initial lump sum payments." (Since the New York City minimums fall below the 3-percent mark for the present, however, no problem exists in this particular case.)

- With respect to the franchise fee ceilings of 3 to 5 percent, and the possibilities of extending fees to include special services, the requirements for a waiver are unclear. Nevertheless, some generalizations are possible. A petition should emphasize heavily any activities that the franchising authority undertakes in relation to cable—e.g., public hearings, publication of notice, conferences with concerned citizens, disposition of subscriber complaints, etc.—in addition to the normal processes of drafting the franchise and monitoring the system's performance. Though the FCC has not addressed the issue, presumably any expenses incurred in the prefranchise process could be included in the franchise fee either as a flat payment or on an amortized basis. In addition, the franchising authority may need to justify requirements of production facilities as "regulatory," as previously discussed. To do so, the franchising authority should keep records as to the facilities' use and perhaps at least initially use them to inform subscribers and nonsubscribers about cable and its regulation.

In light of these uncertainties, what should local franchise authorities do? One recourse is to play it safe and adopt provisions similar to those of the Beverly Hills franchise—but to make clear that annual gross receipts include only those for "regular" subscriber services as discussed above.

Grantee shall pay to the City of Beverly Hills for each calendar year or fraction thereof during the term of this franchise, three (3%) percent of the gross annual receipts from operation of the proposed service; provided, that every such annual payment shall not be less than One Thousand Five Hundred Dollars ($1,500.00).

---

[50] The provisions of the New York City franchises, as well as those in other franchises under which the cable system was carrying broadcast signals prior to March 31, 1972, are grandfathered for a period of 5 years from that date.

In such a case no special showing would be required for the FCC.

Another approach is to waive an annual fee in favor of a relatively small fixed sum plus expenses incurred for granting the franchise—an approach illustrated by the San Mateo franchise.

Fees and Payments. Peninsula [Cable Company] shall pay City the sum of thirty thousand dollars ($30,000) for the rights vested by this franchise. Said payment will be made prior to the exercise of any rights hereby granted and is separate and distinct and will not relieve Peninsula from any payment due City under the terms of this franchise or Article 121 of the San Mateo Ordinance Code, 1959 Edition. Further, Peninsula will reimburse City for all expenses incurred by it in connection with the granting of this franchise, including but not limited to costs incurred by members of the City staff and its consultant, Donald R. Atwell & Associates, Inc. Such payment to be made upon billing and prior to exercising any rights hereunder.

As noted above, a lump-sum payment for a franchising authority's prefranchise expenses presumably would be included in the franchise-fee ceiling.

Another approach is to provide for higher annual fees, in the neighborhood of 5 percent, and to establish a regulatory program that would justify the excess in the eyes of the FCC. This could include the same or higher fee schedules for new services, such as advertising and pay movies. At worst, if the FCC rejects the application, the local franchising authority could set lower, more acceptable fees.

Yet another possibility is to specify a fee of perhaps 5 percent, some or all of which will go into a fund to finance local television programming by community groups. The FCC has expressed its concern about high franchise fees in terms of funds that simply go into the city's general coffers. Its reaction is unclear as to proposals for relatively high fees that are channeled into public service programming or other activities that directly benefit cable subscribers.

Finally, the franchise should include provisions to cover the consequences of the FCC's refusal to authorize a fee. These might take the form of a simple reservation of power clause, which would allow the franchising authority to set a new fee. Or they might specify a particular fee in the event of FCC disapproval. However, the latter approach might undermine the franchising authority's plea for a higher fee and thus invite the FCC to disapprove it.[51]

PROVISIONS FOR TRANSFER OF FRANCHISE

A major issue is the question of sale and of ownership transfer of the cable system. Cable franchises are attractive properties and there is a strong market in them. Nothing in the federal rules speaks directly to what happens when the cable operator sells the franchise. How does the sale affect the certificate of compliance? The issue is particularly important where a community values local ownership or minority ownership or is placing great faith in the cable operator to perform in the field of local programming. Community groups with these interests should make sure that the certificate of compliance places strict conditions on transferability that meet community needs. At the least, the application for the certificate should spell out and the certificate should confirm what procedures will be followed and what

---

[51] For additional discussion of the limitation on franchising fees, see Rivkin, op. cit., Sec. 4.217.

standards applied on the transfer of a certificate. In this respect, the NIMLO Model Ordinance contains useful language:

> The grantee shall not transfer this Franchise to another person without prior approval of the City by ordinance.
>
> In order that the City may exercise its option to take over the facilities and property of the CATV system authorized herein upon expiration or forfeiture of the rights and privileges of the grantee under this Franchise, as is provided for herein, the grantee shall not make, execute, or enter into any deed, deed of trust, mortgage, conditional sales contract, or any loan, lease, pledge, sale, gift or similar agreement concerning any of the facilities and property, real or personal, of the CATV business within prior approval of the City Council upon its determination that the transaction proposed by the grantee will not be inimical to the rights of the City under this Franchise. Provided, however, that this section shall not apply to the disposition of worn out or obsolete facilities or personal property in the normal course of carrying on the CATV business.
>
> Except as provided for in subsection (a) [immediately] above, the grantee shall at all times be the full and complete owner of all facilities and property, real and personal, of the CATV business.
>
> Prior approval of the City Council shall be required where ownership or control of more than 30% of the right of control of grantee is acquired by a person or group of persons acting in concert, none of whom already own or control 30 percent or more of such right of control, singularly or collectively. By its acceptance of this Franchise the grantee specifically grants and agrees that any such acquisition occurring without prior approval of the City Council shall constitute a violation of this Franchise by the grantee.

Thus, the franchise requires the franchising authority's approval for three separate types of transfers: of the franchise, of ownership of tangible property, and of control of the franchisee's corporate entity. Although the franchise touches all major bases, its language might be more specific. First, the prohibition on any "transfer" of the franchise is unclear; it should include any arrangement relating to transfer, such as an agreement to transfer the franchise in the future or a covenant not to seek renewal. Second, and more important, the 30 percent of "control" figure is both too high and too ambiguous. In many public corporations, far less than 30 percent of the stock constitutes effective control; moreover, the franchise does not make clear whether "control" includes devices such as stock options, shareholder agreements, etc. As a result, the franchise should speak in both broad and narrow terms, in order to give the franchising authority maximum flexibility.

The New York City franchises provide an example of more detailed transfer provisions:

> This franchise shall not be assigned or transferred, either in whole or in part, or leased, sublet, or mortgaged in any manner, nor shall title thereto, either legal or equitable, or any right, interest or property therein, pass to or vest in any person, either by the act of the Company or by operation of law, without the consent of the Board. The granting, giving or waiving of any one or more of such consents shall not render unnecessary any subsequent consent or consents.
>
> The consent or approval of the Board to any assignment, lease, transfer, sublease, or mortgage of this franchise shall not constitute a waiver or release of the rights of the City in and to the streets.
>
> The Company shall promptly notify the Board of any actual or proposed change in, or transfer of, or acquisition by any other party of, control of the Company. The word "control" as used herein is not limited to majority stock ownership, but includes actual working control in whatever manner exercised. Every change, transfer or acquisition of control of the Company shall make this franchise subject to cancellation unless and until the Board shall have consented thereto. For the purpose of determining whether it shall consent to such change, transfer or acquisition of control, the Board may inquire into the qualifications of the prospective controlling party, and the Company shall assist the Board in any such inquiry. If the Board does not schedule a hearing on the matter within sixty (60) days after notice of the change or proposed change and the filing of a petition requesting its consent, it shall be deemed to have consented. In the event that the Board adopts a resolution denying its consent and such change, transfer

or acquisition of control has been effected, the Board may cancel this franchise unless control of the Company is restored to its status prior to the change, or to a status acceptable to the Board.

Nothing in this Section shall be deemed to prohibit a mortgage or pledge of the System, or any part thereof, or the leasing by the Company from another person of said System, or part thereof, for financing purposes or otherwise. Any such mortgage, pledge or lease shall be subject and subordinate to the rights of the City under this contract or applicable law.

By considering "actual working control," rather than a set percentage of stock, the New York franchises give the franchising authority considerable flexibility in dealing with corporate changes that have a de facto effect on the system's control. The combination of this broad standard and the NIMLO franchise's specific standard therefore may be advisable.

The Akron franchise includes a much briefer provision for transfer but with the interesting feature that, in no event, shall a transfer take place within 5 years:

The Company shall not assign or in any manner transfer the rights granted by this franchise whether by stock assignment or otherwise in such a manner as to effectively transfer control of such rights except by future consent ordinance of The City of Akron at any time and in no event for a period of five (5) years from the effective date of this ordinance; nor shall these rights be assignable or transferable in any bankruptcy proceedings, trusteeship, receivership, but same shall not be intended to prevent transfer by operation of law by merger or consolidation, and said franchise shall terminate forthwith upon such assignment or transfer.

PERFORMANCE BONDS

To help ensure that the cable operator performs in accordance with the franchise, most franchises include a performance bonding arrangement. In drafting these provisions, the franchisor must realize that performance bonding is not a cure-all. Serious disagreements can arise between the franchisor and the cable operator about the meaning of "adequate performance." The operator may fall behind his construction schedule for reasons that he alleges are beyond his control, but that the franchisor thinks could have been avoided had the operator used better judgment. If the system does not meet technical standards, the cable operator may argue that the fault lies not with him, but with equipment manufacturers; the franchising authority's technical consultants may counter by arguing that the fault lies with improper system design, of which the cable operator should have been aware early in construction. These are but a few examples of the many disagreements that the performance bonding approach cannot resolve.

An example of existing provisions comes from the San Mateo franchise:

Pursuant to Section 121.14 of the San Mateo Ordinance Code, Peninsula will, at the time this franchise is accepted, file its bond in the amount of $200,000. Upon final completion of the system, said bond may be reduced to $20,000.

Besides not specifying standards or procedures for measuring performance, this provision raises a question of what constitutes "final completion" of a system. Systems constantly are being upgraded; a few miles of plant are strung as new housing developments are built or as new sections are added to the franchise area. If the amount of the performance bond is to be reduced after system completion, specific numbers must define completion. For example, a franchise might require that 95 percent of the *existing* homes within the existing franchise area be passed by cable within a period of four years after the franchise is signed.

The NIMLO formulation may be useful in drafting the franchise:

> The grantee shall maintain, and by its acceptance of this Franchise specifically agrees that it will maintain throughout the term of this Franchise a faithful performance bond running to the City, with at least two good and sufficient sureties approved by the City, in the penal sum of $————— conditioned that the grantee shall well and truly observe, fulfill, and perform each term and condition of this Franchise and that in case of any breach of condition of the bond, the amount thereof shall be recoverable from the principal and sureties thereof by the City for all damages proximately resulting from the failure of the grantee to well and faithfully observe and perform any provision of this Franchise.

Again, the NIMLO wording raises the problem of determining what constitutes a breach of the bond's conditions. Moreover, its talk of "damages proximately resulting" from the cable operator's default may create difficulties by importing the notion of proximate cause from an unrelated area of the law; though a cable operator's poor performance may run against the public interest, it may not damage identifiable people or institutions.

Although performance bonds are one useful tool in insuring high-quality service by a cable operator, their administrative difficulties and overkill potential make them an enforcement mechanism that franchising authorities will be slow to use—much like franchise renewals. As a result, a system of easily administered monetary penalties for specific violations may be a useful supplement, along with arbitration procedures as discussed previously.

LIABILITY FOR DAMAGES BY THE CABLE OPERATOR

In addition to performance bonding, cable operators typically must carry liability insurance and agree to indemnify the franchising authority for any legal liability on its part. In this regard the NIMLO Model Ordinance is a useful guide:

> 1. (a) The grantee shall pay and by its acceptance of this Franchise the grantee specifically agrees that it will pay all damages and penalties which the City may legally be required to pay as a result of granting this Franchise. These damages or penalties shall include, but shall not be limited to, damages arising out of copyright infringements and all other damages arising out of the installation, operation, or maintenance of the CATV system authorized herein, whether or not any act or omission complained of is authorized, allowed, or prohibited by this Franchise.
>
> (b) The grantee shall pay and by its acceptance of this franchise specifically agrees that it will pay all expenses incurred by the City in defending itself with regard to all damages and penalties mentioned in subsection (a) above. These expenses shall include all out-of-pocket expenses, such as attorney fees, and shall also include the reasonable value of any services rendered by any employees of the City.
>
> (c) The grantee shall maintain, and by its acceptance of this franchise specifically agrees that it will maintain throughout the term of this Franchise liability insurance insuring the City and ghe grantee with regard to all damages mentioned in subparagraph (a) above in the minimum amounts of (1) $———————for bodily injury or death to any one person, within the limit, however, of $————for bodily injury or death resulting from any one accident; (2) $————for property damage resulting from any one accident; (3) $————for the infringement of copyrights; and (4) $————for all other types of liability.

The Akron franchise provides an example of actual amounts of insurance required:

... The amounts of such insurance against liability due to physical damages to property shall not be less than One Hundred Thousand Dollars ($100,000) as to any one accident and not less than Two Hundred Thousand Dollars ($200,000) aggregate in any single policy year; and against liability due to bodily injury or to death of persons not less than Two Hundred Thousand Dollars ($200,000) as to any one person and no less than Five Hundred Thousand Dollars ($500,000) as to any one accident. The Company shall also carry such insurance as it deems necessary to protect it from all claims under the Workmen's Compensation Laws in effect that may be applicable to the Company. All insurance required by this Agreement shall be and remain in full force and effect for the entire life of this ordinance. Said policy or policies of insurance or a certified copy or copies thereof shall be approved by the Director of Law of said City and be deposited with and kept on file in the office of said Director.

REPORTING REQUIREMENTS

A franchise clearly should specify the kinds of data the cable operator must make available to local authorities. The new FCC rules will make available much important information since they require cable operators not only to file detailed statements with their applications for certificates or compliance, but also to file annual reports about their ownership and programming. Nevertheless, the local franchising authority should reserve the power to require its own reports, since it may need more detailed or more frequent information than the FCC.

The NIMLO Model Ordinance provides guidance, but deals almost exclusively with the filing of financial accounts:

The grantee shall file with the City Clerk true and accurate maps or plots of all existing and proposed installations.

The grantee shall file annually with the City not later than sixty days after the end of the grantee's fiscal year, a copy of its report to its stockholders (if it prepares such a report), an income statement applicable to its operations during the preceding 12 months period, a balance sheet, and a statement of its properties devoted to CATV operations by categories, giving its investment in such properties on the basis of original cost less applicable depreciation. These reports shall be prepared or approved by a certified public accountant and there shall be submitted along with them such other reasonable information as the City Council shall request with respect to the grantee's properties and expenses related to its CATV operations within the City.

The grantee shall keep on file with the City a current list of its shareholders and bondholders.

The ACLU Model Code goes further in requiring extensive financial disclosure and reporting of action taken regarding customer complaints:

Each franchisee shall file with the *Commission* quarterly reports signed by a Certified Public Accountant, on gross revenues. Each franchisee shall also allow its franchisor and the *Commission* to audit all of its accounting and financial records upon reasonable notice; it shall also make available all of its plans, contracts, and engineering, statistical, customer and service records relating to its system and to all other records required to be kept hereunder. Each franchisee shall at all times maintain complete and accurate books of account records of its business and operations, and all other records required by this Code. In addition, each grantee shall file annually with the *Commission* and its franchisor an Ownership report, indicating all persons who at any time during the preceding year did control or benefit from an interest in the franchise of more than 1%, and all creditors, secured and unsecured, in excess of $1,000. The report shall also include all creditors whose accounts were at any time paid in full for a period of greater than two months, together with the reasons therefor. Each grantee shall also file annually with the *Commission* and its franchisor copies of such rules, regulations, terms and conditions which it has adopted for the conduct of its business.

All user and other complaints sent to the franchisee, including those forwarded by government agencies, shall be turned over to the *Commission* and to the franchisor within ten days of mailing, together with the franchise's response thereto. All complaints to the *Commission* or franchisor shall be forwarded to the appropriate grantee and answered by each within 10 days of original mailing.

Each grantee shall file with the *Commission* a quarterly report of all complaints and trouble calls received, and shall include in that report a breakdown of the calls by category, the number of second or subsequent calls on the same complaint, and an average of the period of time required to satisfy each complaint reported.

FRANCHISOR'S RIGHTS

The franchise should include explicit statements of the franchisor's right to inspect the cable system, review the operator's books, and adopt additional provisions. The MVCOG model franchise provides a number of interesting provisions in these areas:

The franchisor hereby reserves the right to adopt, in addition to the provisions contained herein and existing applicable ordinances, such additional regulations as it shall find necessary in the exercise of its police power; provided, however, that such regulations, by ordinance or otherwise as provided by law, shall be reasonable and not in conflict with the rights herein granted.

The franchisor shall have the right to inspect the books, records, maps, plans, income tax returns, and other like material of the grantee at any time during normal business hours.

The franchisor shall have the right during the life of this franchise to install and maintain, free of charge, upon the poles of the grantee any wire and pole fixtures necessary for a police alarm system, fire alarm system, traffic control system, or other similar system or systems for other governmental functions, on the condition that such wire and pole fixtures do not interfere with the cable television operations of the grantee.

The franchisor shall have the right to supervise all construction or installation work performed under the provisions of this franchise and to make such inspections as it shall find necessary to ensure compliance with the terms of this franchise and other pertinent provisions of law.

At the expiration of the term for which this franchise is granted, or upon its termination and cancellation as provided for herein, the franchisor shall have the right to require the grantee to remove at its own expense all portions of the cable television system from all public ways within the franchise area.

CONCENTRATION OF CONTROL

Some franchises prohibit cable operators from leasing, selling, or maintaining television sets. To an extent this reflects pressures from local retailers and repairmen, who fear that cable operators will pose a competitive threat. It also reflects the view that the cable operator's special access to subscribers creates an unfair and quasi-monopolistic competitive element. The Akron franchise illustrates such provisions:

The Company agrees to restrict its operation within the City so as not to compete with the television sales, service and repair industry; that is, it shall not offer nor accept employment directly or indirectly in the repair or servicing of a customer's television set or sets other than the technical servicing that may be needed in the cable installation within the home and its connection to the customer's television set. Nor will the Company engage directly or indirectly in the referral of such repair or servicing to any particular repair or service agency.

The Illinois Commerce Commission emphasizes the alleged unfair competitive aspects of permitting the cable operator to be involved in set sales and leasing:

Most franchises now prohibit cable systems and their affiliates, shareholders, officers and directors from engaging within their franchise areas in the sale, rental or repair of television receivers. This is designed to prevent unfair competitive access to the home, is eminently reasonable, and would appear suitable for adoption by the Commission as its own rule.[52]

An earlier Rand study, however, suggests that permitting cable operators to lease sets may reduce their maintenance cost and also stimulate the production of sets specially designed for cable.[53]

Recognizing this possibility, the MVCOG franchise prohibits cable operators from engaging in television set leasing, sales, and repair in the short term, but leaves the door open for developing integrated services in the long term:

(a) The grantee shall not engage in the business of selling, repairing, or installing television receivers, radio receivers, or accessories for such receivers within the franchise area.

(b) Failure of the grantee to honor its obligation under subsection (a) shall be a material breach of this franchise.

(c) The provision in subsection (a) above may be renegotiated after five years at the option of the franchisor to allow the grantee the non-exclusive right to sell, lease, and service television receivers.

Some local franchising authorities also may be concerned with preventing a cable system from becoming a minor cog in the wheels of a giant conglomerate communications corporation. These fears appear to be justified, since the trend toward cross-ownership and common ownership of cable systems is strong. At least fifty percent of existing cable systems are owned by broadcasting or publishing interests. This tendency probably stems from the existing media's desire not only to neutralize possible competition, but also to share cable's potential profits. As a result, a local franchising authority may wish to prohibit cross- or common ownership of a local cable system. The New York City franchises provide a good example:

Neither the Company nor any officer or director of the Company shall hold, directly or indirectly, any stock or other beneficial ownership interest in any other company owning or operating: a System within the City; any radio or television broadcast station whose signals are carried on the System on a regular basis; any television broadcast network other than a network consisting entirely or substantially of community antenna television systems; or any newspaper or magazine whose principal circulation market is New York City, except that ownership by an officer or director of less than one percent (1%) of the outstanding stock of any company whose securities are listed or admitted to trading on a national securities exchange shall not be deemed a violation of this Section. No officer or director of the Company shall be an officer or director of any company owning or operating businesses of the types heretofore mentioned.

So far, the FCC has taken relatively little action to prevent concentration of control, by imposing bans on cable ownership only by networks, local television stations, and telephone companies. It is unclear whether the FCC's rules preempt local governments in adopting more extensive cross-ownership restrictions. As a result, provisions like the above may require FCC approval.

MAINTENANCE OF HOME ANTENNAS

Some franchises require cable operators to give subscribers the option of easily switching back to their own antennas. This puts the cable operator under additional

---

[52] *Notice of Inquiry,* p. 41.
[53] See Johnson et al., op. cit., Paper 9, pp. 51-61.

pressures to give good service. The Illinois Commerce Commission provides useful guidelines in this respect:

> The Commission believes that subscribers who wish to retain their own antennas for television broadcasts should be free to do so. Operators will therefore be required to provide, on request and without additional charge, a switching device allowing a subscriber to use his own television antenna as he chooses. Operators will not be permitted to require the removal, nor offer to remove, any existing antenna as a condition of providing cable service.[54]

EMERGENCY USE

Language drawn from the New York City franchises and from the NIMLO Model Ordinance, respectively, provides useful guides covering the use of cable systems in emergencies:

> In the event of an emergency situation, as determined by the Director of Communications, the City may interrupt signals otherwise being distributed by the Company for the delivery of signals necessitated by such emergency.

> In the case of any emergency or disaster, the grantee shall, upon request of the franchisor make available its facilities for emergency use during the emergency or disaster period.

SEPARABILITY OF CLAUSES, COMPLIANCE WITH APPLICABLE LAWS

As noted before in relation to several particular issues, the franchising authority should reserve rulemaking power to cope with new developments. The MVCOG model franchise provides good examples of miscellaneous legal provisions and of methods for modifying the franchise in accordance with future FCC regulations.

> If any section, subsection, sentence, clause, phrase, or portion of this franchise is for any reason held invalid or unconstitutional by any court of competent jurisdiction, such portion shall be deemed a separate, distinct, and independent provision and such holdings shall not affect the continued effectiveness or validity of the remaining portions hereof.

> At all times during the life of this franchise, the grantee shall be subject to all lawful exercise of the police power by the franchisor and to such reasonable regulation as the franchisor shall hereafter provide.

> In addition to any provision of this franchise, and notwithstanding any provision thereof, the grantee shall comply with all applicable state and federal laws and with all applicable regulations of the Federal Communications Commission.

> This franchise shall incorporate any modifications made by the Federal Communications Commission in its regulations concerning franchise standards, if such incorporation is required by said regulations. Should such modification require a substantive provision not delineated in the regulations, the grantee and the franchisor shall negotiate such additional term. If agreement is not reached, the provision shall be determined in accordance with the procedure in Section 25(f).

[54] *Notice of Inquiry,* p. 30.

RECEIVERSHIP

Although many franchises do not cover this area, the franchising authority should take into account the contingencies of receivership, reorganization, and bankruptcy. The language of the New York City franchises may provide a useful guide:

The Board shall have the right to cancel this franchise one hundred and twenty (120) days after the appointment of a receiver, or trustee, to take over and conduct the business of the Company, whether in receivership, reorganization, bankruptcy, or other action or proceeding, unless such receivership or trusteeship shall have been vacated prior to the expiration of said one hundred and twenty (120) days, or unless:

1. within one hundred and twenty (120) days after his election or appointment, such receiver or trustee shall have fully complied with all the provisions of this contract and remedied all defaults thereunder; and,

2. such receiver or trustee, within said one hundred and twenty (120) days shall have executed an agreement, duly approved by the court having jurisdiction in the premises, whereby such receiver or trustee assumes and agrees to be bound by each and every provision of this contract.

Though giving the receiver a chance to remedy the system's defaults seems fair, this provision's lack of standards may create conflict as to whether or not the receiver has succeeded; moreover, 120 days may be too short time in which to cure an already ailing system.

CANCELLATION AND EXPIRATION

The franchise clearly should provide for cancellation and expiration of the franchise, including terms under which the system is to be purchased by the city or by another grantee. The New York City franchises go into particularly great detail:

(a) The Board shall have the right to cancel this franchise if the Company fails to comply with any material and substantial provision of this contract, or any reasonable order, direction or permit issued by any City agency pursuant to such material and substantial provision, or any rule or regulation promulgated by the Director of Franchises which is reasonable in light of, and consistent with, any provision of this contract; or if the Company persistently fails to comply with any provision of this contract, or any reasonable order, direction or permit issued by any City agency pursuant to any provision of this contract. Such cancellation shall be by resolution of the Board duly adopted in accordance with the following procedures:

1. The Director of Franchises shall notify the Company of the alleged failure or persistent failure of compliance and give the Company a reasonable opportunity to correct such failure or persistent failure or to present facts and argument in refutation of the alleged failure or persistent failure.

2. If the Director of Franchises then concludes that there is a basis for cancellation of the franchise pursuant to this subdivision (a), he shall notify the Company thereof.

3. If within a reasonable time the Company does not remedy and/or put an end to the alleged failure or persistent failure the Board, after a public hearing on notice, may cancel the franchise if it determines that such action is warranted under this subdivision (a).

(b) If for ten (10) consecutive days the System, or any part thereof, is inoperative, or if the same is inoperative for thirty (30) days out of any consecutive twelve (12) months, the Board may cancel this franchise.

(c)　The Company shall not be declared in default or be subject to any sanction, under any provision of this contract in any case in which the performance of any such provision is prevented for reasons beyond its control.

(d)　If all or any part of the streets within the District are closed or discontinued as provided by statute, then this franchise, and all rights and privileges hereunder with respect to said streets or any part thereof so closed or discontinued, shall cease and determine upon the date of the adoption of the map closing and discontinuing such streets, and the Company shall not be entitled to damages from the City due to the closing or discontinuance of such streets or for injury to any part of the System in the streets or for the removal or relocation of the same.

(e)　If the System is taken or condemned pursuant to law, this franchise shall, at the option of the Board, cease and determine on the date of the vesting title pursuant to such taking or condemnation, and any award to the Company in connection with such taking or condemnation shall not include any valuation based on this franchise.

(f)　Upon cancellation or expiration of this franchise, the City shall have the right to purchase the System in accordance with subdivision (g) of this Section, and the Board may direct the Company to cease operation of the System. If the City elects to purchase the System, the Company shall promptly execute all appropriate documents to transfer title to the City, and shall assign all other contracts, leases, licenses, permits and any other rights necessary to maintain continuity of service to the public. The Company shall cooperate with the City, or with another person authorized or directed by the Board to operate the System for a temporary period, in maintaining continuity of service. Nothing herein is intended as a waiver of any other rights the City may have.

(g)　If this franchise;

    1.　is cancelled by the Board by reason of the Company's default, that part of the System located in the streets shall, at the election of the City, become the property of the City without any charge therefor; that part of the System not located in the streets shall, at the election of the City become the property of the City at a cost not to exceed its then book value (i.e., cost less accumulated depreciation) according to generally accepted accounting principles, with a reduction for any damages incurred by the City in connection with such cancellation. Such book value if not agreed upon, shall be determined by arbitration pursuant to Section 20 of this contract, but shall not include any valuation based upon this franchise. Damages incurred by the City shall include without limitation, any payments made by the City pursuant to a resolution of the Board authorizing or directing another person to operate the System for a temporary period until a franchise therefor is granted.

    2.　terminates by expiration of its term, the purchase price to the City for the System shall be its then fair value as determined by arbitration held pursuant to Section 20 of this contract. Beginning within two years prior to expiration and whether or not the City has then elected to purchase the System, either the City or the Company may demand an arbitration pursuant to Section 20 of this contract, for the purpose of determining fair value of the System on the date arbitration was demanded which determination shall be subject to correction or adjustment by the arbitrators to reflect the fair value on date of expiration, to be paid by the City if it elects to purchase the System. Such fair value shall be the fair value of all tangible and intangible property forming part of the System but shall not include any valuation based upon this franchise. If the City does not purchase the System, the Company shall remove that part of the System located in the streets and restore the streets to a condition satisfactory to the Commissioner of Highways.

(h)　Upon the cancellation by the Board, or upon the expiration, of any other franchise to construct, maintain and operate a broadband communications facility, the Board may, by resolution, direct the Company to operate the same for the account of the City for a period of six (6) months and the Company agrees to comply with such direction. The City shall pay the Company all reasonable and necessary costs incurred by it in operating such broadband communication facility.

Though comprehensive in their scope, the New York City provisions illustrate difficulties as to both the procedures and terms of franchise termination. First, the standards for franchise termination (such as what does and does not constitute "default") are necessarily a matter of interpretation. They cannot be made significantly more specific by any reach of the draftsman's pen. As a result, the fairness of any franchise termination will depend largely on the good faith of the franchising authority and of the cable operator.

Second, Subsection g's provisions for compensation of the operator upon franchise termination are severe. If the franchising authority "cancels" the franchise "by reason of the Company's default," the cable operator forfeits his trunk, feeder lines, and droplines "in the streets" and receives no more than "book value" for his other equipment. A cable system's largest investment is in its lines, and the book value of other equipment probably will be quite low, since most cable operators use accelerated depreciation practices. As a result, cancellation may amount to virtual confiscation—a remedy so severe that a franchising authority would be most reluctant to invoke it. Similarly, the system's book value—as opposed to fair market value —also is likely to be low, once again putting the franchising authority under pressure not to cancel the franchise.

One alternative approach would be to adopt a "fair market value" standard. A second would be to set a purchase price at the very beginning in the franchise itself, such as a given percentage of the purchase price of the system's equipment and with the percentage adjusted in accordance with the equipment's age. A third would be to offer the system to the highest bidder acceptable to the franchising authority, with all proceeds going to the original franchisee. These approaches would not only give the cable operator more security, but also provide the franchising authority with a pragmatically usable enforcement mechanism.

IV. CONCLUDING REMARKS

As even the most cursory reading of this report indicates, the franchising process is complex, difficult, and frustrating. Nevertheless, local governments and community groups must plunge into it deeply and immediately. Well-conceived decision-making is important at this early stage of the industry's development. Since hindsight is clearer than foresight, this report does not begin to cover the myriad problems yet to be discovered, let alone resolved. Nevertheless, local franchising authorities must wield a better-informed and more sophisticated hand now than they did in the past.

In pursuing this process, franchising authorities must keep carefully in mind that in certain areas the FCC has constrained the range of local decisionmaking. In particular, the FCC has imposed strong requirements with respect to maximum franchise fees, duration of franchise, broadcast signals to be carried, minimum channel capacities, and allocation of free channels. At the same time, local franchising authorities have wide latitude in setting rates for services, specifying the facilities for local program origination, and defining geographic boundaries within which one or more cable systems are to operate. Within the range of choice open to franchising authorities, many hard decisions remain.

Moreover, local authorities will need to pay attention to activities at the state level. An increasing number of states are expressing interest in regulating cable to one degree or another. Some states have preempted functions that otherwise would be undertaken at the local level. Regardless of how these activities are split between state and local jurisdictions, the basic considerations discussed in this report will be of paramount concern to whoever is to bear the responsibilities, at either the local or state level, for discharging these tasks.

Appendix A

CHECKLIST OF MAJOR ELEMENTS IN THE FRANCHISING PROCESS

This checklist serves not only as a short summary of this report, but also as an easy means for local governments and citizen groups to evaluate their franchising process. Two caveats are in order, however. First, by its very nature this checklist is highly abbreviated. Any real analysis of an issue requires reading the relevant portion of this report as well as other materials. Second, not all the steps and standards noted here will be appropriate for all localities. This report is intended not as a master plan for all communities, but rather an exposition of major considerations and alternatives. Each community must choose its own steps and standards.

I. The Process of Franchising

- To select a cable operator, has the franchising authority decided to use:
  - the "negotiation" approach?
  - the "competitive bid and award" approach?
  - a variant of the two?

- Has the franchising authority adopted a detailed procedural framework to govern the drafting and award of the franchise?

- Does this procedural framework require that at all major decisional points the franchising authority:
  - hold meaningful public hearings?
  - give effective public notice?
  - publish a written opinion explaining any decision it reaches?

- Has the franchising authority considered multiple versus single system ownership?

- Has the franchising authority used some means—e.g., Council of Governments, regional authority, state agency—to communicate with neighboring jurisdictions concerning their franchising policies?

147

- Has the franchising authority considered the advantages and disadvantages of commercial, noncommercial, mixed commercial and noncommercial, and municipal system ownership?

- Has the franchising authority held hearings on and adopted a draft franchise that indicates which provisions are open to bids?

- Has the franchising authority drafted and disseminated a request for proposals (RFP) as widely as possible?

- Does the RFP require full disclosure of the applicant's financial, ownership, character, technical, and other qualifications?

- Has the franchising authority independently analyzed the area's economic potential for cable and compared all proposals with its findings?

- Has the franchising authority analyzed each applicant's financial ability to deliver its proposed service?

- Has the franchising authority given effective public notice, held meaningful public hearings, and published a written opinion in selecting an applicant for the final franchise award?

- Is the franchising authority prepared to:
  — join the cable operator in making any necessary special showings to the FCC?
  — oppose the cable operator's application for a certificate of compliance from the FCC?

II. Contents of the Franchise

- Do the prefatory statements state compliance with federal and state law?

- Are the definitions sufficiently precise and comprehensive?

- Does the franchise comply with the FCC's fifteen-year maximum duration for initial franchises and a "reasonable" period for renewal franchises?

- Is the franchise nonexclusive in terms of geographic coverage?

- Does the franchise give the franchising authority some control over the cable system's choice of broadcast television signals?

- Does the franchise require the cable operator to construct the system speedily and equitably?

- Does the franchise specify how closely a trunk line must pass each dwelling?

- Does the franchise specify standards as to:
  — underground versus aboveground wiring?
  — quality of underground and aboveground construction?
  — protection of property owners' rights?

- Does the franchise include forceful but workable prohibitions against discrimination in employment practices?

- Does the franchise include technical standards that are specific enough to be enforceable?

- Does the franchise require the cable operator to resolve subscribers' service complaints efficiently and expeditiously?

- Does the franchise include a provision that insures access to apartment houses by the cable operator and protects landlords from property damage?

- Does the franchise provide for:
  — rate-of-return regulation?
  — rate surveillance?
  — no rate control at all?

- Do the franchise's rates or rate control mechanism take into account:
  — differences in underlying costs?
  — power to set rates for new services?
  — periodic rate modifications?

- Do initial installation rates reasonably reflect the cable operator's cost for different subscribers?

- Does the franchise provide for disconnection and reconnection charges?

- Do monthly subscriber service rates reflect:
  — rate differences in serving different subscribers?
  — rate differences for additional outlets in the same home?
  — rate differences in providing TV and FM reception?

- Does the franchise set charges for relocation of outlets within the same dwelling unit?

- Does the franchise specify subscriber billing procedures?

- Does the franchise limit the cable operator's ability to offer reduced rates?

- Does the franchise require more free or preferentially priced channels than the FCC's rules allow?

- Does the franchise prohibit or unduly restrict pay television?

- Does the franchise set a minimum channel capacity?

- Does the franchise provide for interconnection of the cable system with other neighboring systems?

- Does the franchise provide for a fee in excess of the FCC's limitations?

- Does the franchise restrict transfer of the franchise or the cable system?

- Does the franchise include a clear and workable performance bond?

- Does the franchise require the cable operator to disclose necessary information to the franchising authority?

- Does the franchise restrict cross- or common ownership of the cable system?

- Does the franchise require the cable operator to allow a subscriber to switch back to his own antenna easily?

- Does the franchise provide that:
  — its provisions are severable?
  — the cable operator will comply with all laws?

- Does the franchise make provision for receivership of the cable system?

- Does the franchise specify fair and realistic standards and procedures for cancellation of the franchise?

Appendix B

SINGLE VERSUS MULTIPLE OWNERSHIP

Does it make much difference whether one or several operators serve the franchise area? This question will be treated with respect to the following major factors:

- Technical capability and interconnection;
- Division of geographic coverage;
- Sharing of program origination facilities;
- Economies of scale;
- Local control;
- Yardsticks for comparing performance;
- Investment requirements and construction schedules;
- Satisfying rival claimants.

TECHNICAL COMPATIBILITY AND INTERCONNECTION

One potential problem of multiple ownership is the difficulty of maintaining adequate compatibility and interconnection among the districts. With separate owners—including private, municipal, or community nonprofit organizations—attaining these goals will not be easy. Some cable operators may install only one cable and put a converter in each home in order to provide 20 to 24 channels. Others may opt for a dual-cable system with a converter in order to provide 40 or so channels. Yet the extent to which a community benefits from a large number of channels in one district will depend on the offerings in other districts. One district's 40-channel capacity will be of limited value if other systems have only 20. For example, colleges and universities may need 5 channels for home instruction, but may have enough capacity only in the 40-channel districts. The benefits of televised instruction would be reduced to the extent that students in 20-channel districts could not be reached. A special continuing medical education channel may be economically feasible only if it covers the entire metropolitan area, but sufficient capacity may exist only in the 40-channel districts.

If this interrelation exists, the separate owners might not necessarily enter a

151

voluntary agreement to ensure capability and interconnection. First, cable operators may have honest disagreements about how much capacity is needed and what technology is most effective. One operator may conclude that the 20 channels are enough, since this quantity would meet the FCC's minimum channel requirements.[1] Others may be more optimistic about services and favor 40 channels. These disagreements are apt to be particularly severe if the type of ownership—e.g., public versus private—varies from one district to another. This is not to say that the United States cable industry should be limited to one technology and channel capacity. Diverse technologies (single cable versus dual cable, use of converters, and experiments with switched systems) clearly need to be pursued. Yet within any one jurisdiction, common technical characteristics are essential for the development of services where broad coverage is required. Diversity is best achieved *among,* not within, separate franchising jurisdictions.

Obtaining voluntary agreement will be troublesome to the extent that cable operators are interested primarily in retransmitting broadcasting signals—as most of the industry is today. For this use 20 channels are more than adequate. Moreover, the cable operator may reason that he can lease additional channels at little or no more than their incremental cost, thus making additional profits zero or very low. He therefore may opt for 20 channels rather than 40. Other operators may be concerned about the longer-term value of additional channel capacity and thus may disagree. Their attempts to provide large capacities may lead to frustration, however, because their additional channels cannot provide services to other districts.

Moreover, unless each operator is bound by a carefully devised and enforced plan for interconnection, he may design his system suboptimally to cover only his district. For example, he may locate his headend at a point that is convenient for his own district, but that has no line-of-sight path for microwave interconnection to other headends. This possibility becomes all the more probable if separately operated systems follow different construction schedules. One operator may complete his headend and much of his plant before a neighboring operator has even decided where to locate his headend. Again, cable operators theoretically could arrive at a voluntary agreement to coordinate construction schedules and interconnection, but this is a difficult task. Cable operators primarily interested in broadcast signals may decide to get these signals off the air by microwave links and hence conclude that they do not need interconnection. They may reason that interconnection would be important only for educational, governmental, and other uses and that these uses would generate little or no additional profits.

To be sure, these problems are difficult, but not insurmountable. If all districts' franchises require common technical standards, construction schedules, and other conditions essential for coordination, the outcome could be the same as that under common ownership. In addition, state cable regulation is increasingly common and may create uniformity as well as coordination. As a result, single and multiple ownership of cable systems within a franchise area can have similar results, depending upon the regulatory effort invested.

---

[1] For a fuller discussion, see Rivkin, op. cit.

DIVISION OF GEOGRAPHIC COVERAGE

Another problem of divided ownership is determining the geographic boundaries of each district. One easily can visualize disputes among cable operators over serving particular subareas in the franchise area. In seeking to enlarge their potential subscriber base, some operators may encroach on others' territories. Thus, separate ownership may create serious difficulties in dividing up the separate districts.[2]

One potential solution is not defining boundaries at all: that is, franchising several cable operators to serve any areas they choose. Each would serve the area where he could build a plant more quickly than the other operators. This would have the added advantage of encouraging speedy coverage of as large an area as possible. It would be an attractive approach if the retransmission of broadcast signals characteristic of today's cable industry were the only concern. More advanced services, however, may require that cable systems follow boundaries of school districts or particular communities of interest—such as ethnic, industrial, governmental, or commercial lines.

Geographic division is not an insuperable problem, but will entail extended debate and disagreement to reach final decisions—a process that could be avoided through single ownership.

ECONOMIES OF SCALE

Substantial economies of scale are possible for cable systems that offer advanced services requiring central computers, local origination facilities, and microwave interconnection. Unless each district encompasses enough dwellings to exploit most of the economies of scale, however, the separate owners must agree to share the cost of central computers and other facilities serving the entire area. So far as maintenance and other operating costs are concerned, there would be little difference between separate and single ownership for districts of 20,000 subscribers or more; crossing of boundaries by maintenance crews might be marginally easier under common ownership than under separate ownership.

SHARING OF FACILITIES FOR LOCAL PROGRAM ORIGINATION

Whether they have single or multiple ownership, larger cities—those with populations of half a million or more—probably will need several separate local program origination facilities scattered about the city to serve local community needs. It would also be advisable to have one central facility with relatively elaborate studios and equipment, however, to distribute higher-quality programming for

---

[2] An example of this problem arose in the franchising of Las Vegas, where the division of the city between two franchise holders was accomplished by voluntary bargaining between the two. After a long period of argument and debate they finally agreed simply to split the city along the Strip. See Mitchell, op. cit., p. 41.

the entire area. Splitting the metropolitan area into separate districts might create problems of sharing the cost of a common facility. Some cable operators might argue that the cost simply should be split equal ways. Others might argue that their share of the cost should depend on the extent to which they carry programming originating at the central facility. Yet others might maintain that costs should be divided in proportion to the number of subscribers served by each district.

Again, this problem is not insuperable, but it could seriously delay design and construction of the cable system. Multiple franchises, then, will require special attention from the franchising authority.

LOCAL CONTROL

One frequently mentioned potential advantage of separate ownership is a greater degree of local participation in controlling channel use and in setting monthly subscriber rates. If each district is designed to be economically self-sufficient, rates would vary from district to district in accordance with underlying service costs. These costs depend on each district's capital investment and operating expenditures relative to the number of users. Generally, the greater the population density, the lower the cost per home passed by cable; the higher the level of cable penetration, the lower the cost per subscriber. Rates reflecting these costs would prevent cross-subsidization among districts. Thus some subscribers would be better off and others worse off with separate rates for the separate districts.

At the same time, single ownership for the entire area does not foreclose rate variation among districts. The franchise could specify separate rates for each geographic area, depending on which cable headend serves it.

The single-ownership approach has flexibility because it leaves open the *option* of having either a single rate for the whole area or different rates for separate districts and subdistricts. The separate-ownership approach might lock the system into separate rates to the extent that subscriber costs vary among the districts and each district must be economically self-sufficient. On the other hand, under single ownership a franchising authority would be free to set uniform rates for all districts, regardless of differences in underlying costs, so long as total revenues do not fall below total costs.

If the franchisor set a uniform rate structure for the whole area, some districts might not be economically self-sufficient and would require subsidization. Transferring funds from one district to another would be more difficult with separate ownership than with single ownership. Conversely, a uniform rate structure which made even the least profitable districts viable would result in windfalls for operators in more profitable districts.

Another aspect of local control is access to local cable channels. Some groups feel that local ownership will give them access to channels under more favorable terms than will metropolitan-wide system ownership—i.e., that they can influence a local operator more easily than they can a city-wide operator. They may be right. A powerful community organization indeed may have greater leverage in gaining access to a relatively small locally owned system than to a larger system "run" by a large organization in another part of town. Yet, a serious question arises as to whether access to cable channels ought to depend on community groups' power and influence. It may be more important to require enough channel capacity to provide all groups with equal and nondiscriminatory access regardless of the pressure— political and otherwise—that they are able to exert.

YARDSTICKS FOR COMPARING PERFORMANCE

One clear advantage of the separate-ownership approach is that it enables comparisons of performance among cable operators. If one cable operator is doing poorly, his shortcomings will stand out more clearly if there is a neighboring cable operator who is doing well. These performance comparisons may be especially useful in deleting, modifying, and adding provisions to franchise renewals.

INVESTMENT REQUIREMENTS AND CONSTRUCTION SCHEDULES

Another advantage of separate ownership is that it substantially reduces the investment requirement for individual operators. For example, the overall metropolitan system in the Dayton, Ohio area is estimated to involve an investment of about $22.5 million. For today's cable industry, this is a large amount of money for a single operator to raise. The cable industry's increasing trend toward merger, however, will make funding of this magnitude progressively easier.

More important than the total investment is the question of whether an area's market for cable services is large enough to make the enterprise economically viable. If it is, then funding probably would be available under either single or multiple ownership.

Related to the question of capital requirements is the issue of construction timetables. Construction probably could be completed sooner in separately franchised districts than in a single-ownership system. Again, much depends on whether the single owner could obtain funding as easily as the separate operators in each of the districts. To the extent that individual operators could obtain funds more quickly, they also could complete the detailed engineering planning, purchase materials, and hire construction crews more quickly than could a single operator.

SATISFYING RIVAL APPLICANTS

One advantage of multiple franchises is simply that they permit more applicants to be satisfied. On strictly political grounds, a franchising authority may prefer approving 3 out of 10 applicants to selecting only one. While each would have a smaller portion of the pie than would a single approved applicant, the franchising authority may face less political pressure.

Chapter 3

Technical Considerations in Franchising

Carl Pilnick

I. INTRODUCTION

The long-awaited Federal Communications Commission (FCC) regulations on cable television, which became effective on March 31, 1972, attempted to bring some degree of order to the relatively chaotic process of granting cable television franchises. There are still many problems in the franchising process, however, particularly in the large metropolitan regions of the United States, where the FCC rules are intended to assist the penetration of cable television.

Some of the factors that complicate the task of developing an appropriate franchise are the following:

- Cable television is in a transition stage, at least for the major markets. The urban cable system, unlike its rural counterpart, will be used for a variety of purposes, some completely commercial and others involving public services. The future mix is almost impossible to forecast at present. Consequently, some decisions that ideally would be part of the franchising process cannot be made intelligently at this time.
- The market demand and economic feasibility for many projected new cable services are yet to be demonstrated. There is a "chicken and egg" syndrome in which many new services are feasible only with a large subscriber base. This requires early expansion of the physical cable plant, but premature expansion may simultaneously constrain some desirable services. As an example, a community fire-alarm system using the cable to transmit alarms to a central point would not be useful unless it included almost all subscribers. In the normal course of events, however, the cable system would be designed and installed some years before such a fire-alarm system could be planned. The cable design may or may not later accommodate such a service, and, in effect, might inhibit the fire-alarm system from becoming a reality.
- Cable technology is developing rapidly; components and techniques in use today probably will have a high obsolescence rate.
- Technical standards imposed by the FCC are limited in scope and apply only to broadcast TV programs distributed through the cable system. No standards are established for cablecasting or for channels that will carry new services. No attention has been paid to related characteristics such as system *reliability*, which is important for many new services.
- Franchises enacted in the past are proving deficient in the areas of moni-

toring technical performance and assuring compliance with even the minimal standards established. Procedures for determining compliance are inadequate in many cases, and in others there are no proportionate corrective or penalty provisions. The threat of disenfranchisement is the ultimate penalty, of course, but it is an "over-kill" solution and simply not realistic enough to cope with a wide variety of technical deficiencies.

- Many city or county officials involved in the franchising process believe that the standard, entertainment-oriented, cable system design can provide a host of new communication services at little or no extra cost. In fact, many new services are not compatible with conventional cable system design. To achieve these, a higher initial system cost may be required, a cost that subscriber revenues alone cannot cover.

- The cable system operators, faced with ambiguous regulations and a lack of definition of exactly what capability their systems should have for the future, naturally will lean toward a design concept that minimizes initial system cost. This is usually achieved at the expense of flexibility for expansion.

Factors such as these result in a franchising dilemma. In a system as complex as a cable communication network, the technological, economic, and regulatory parameters are interdependent; none can be determined without full consideration of the others.

To clarify what this means and to assess the implications, it is useful to focus on how the cable system will be used. For the major markets, at least, cable television system is now a hybrid form, consisting of three elements:

1. *An entertainment distribution network* fulfilling the classical, purely commercial, community-antenna television (CATV) function of receiving and retransmitting broadcast television programs.

2. *A quasi-public communications system* serving at least the local government and the schools. By FCC fiat, if for no other reason, the franchising city or county must concern itself with at least two cable channels, the local government and the education channels. (A third channel for public access is presently the responsibility of the cable-system operator, but its function appears so closely related to the public interest that it is inconceivable the local government can remain completely divorced.) These two services require an immediate involvement by the franchising authority; this involvement is expected to grow in the future.

3. *A transmission medium for a variety of new "nonbroadcast" services,* which might, in the future, include remote shopping, pay TV, fire or burglar alarm systems, and a host of other possibilities. Each new service would presumably be introduced when feasible and economically attractive, and its application could be classified anywhere between completely commercial and completely public-service. Pay TV would be an example of the former category; two examples of the latter might be a fire-alarm system in which sensor-activated alarms are transmitted through the cable, and a school-to-school television link. An "in-between" application might be that of a cable channel leased by a publicly supported university in order to provide adult extension courses to subscribers, for a fee.

This third category is particularly difficult to grapple with in terms of the franchise. Until the new services move from the "potential" into the "actual" category, the exact nature and degree of involvement by the franchising authority cannot be precisely determined. During the life of a cable TV franchise, some 10 to 15 years, the city's participation can certainly be expected to increase dramatically.

The cable network configuration and equipment that fulfill the requirements of points 1 and 2 above may not be easily adaptable to many of the newer nonbroadcast services. A "minimum-initial-cost" cable-system design may place excessive emphasis on present usage at the expense of services that may well prove to be more important in the future.

On the other hand, it is unreasonable to demand a system design that places disproportionate priority on future services that are only possibilities. The franchise holder quite legitimately fears being compelled to incur costs in the initial cable-system installation for which there may never be any counterbalancing revenues. Certainly, if such costs seriously jeopardize the profitability of the venture, neither the city, the system's subscribers, nor the system operators will gain.

A key objective in the franchising process, therefore, is to strike a balance between the two extremes and, in effect, answer the question of how a franchise can best be written to assure:

- Sufficient flexibility in both system design and franchise structure so as not to foreclose too many options for future expansion, and at the same time not impose an unnecessary financial burden on the franchisee;
- The inclusion of adequate technical standards, procedures for determining compliance, and appropriate corrective or penalty provisions;
- Expeditious construction of the "initial service" portion of the system;
- The franchising authority's ability to adjust appropriately the extent of its participation in future specific services at the time the services become a reality.

This report suggests a procedure that relates cable-system technical design to the franchising decisions in a more explicit way. The approach carries with it some penalties (primarily in the need for earlier system planning and an increase in initial system cost), but, if all goes well, will result in a cable-system evolution that is more responsive to future needs.

The key features of the approach are:

- To specify a flexible initial design concept that will reduce technological risk when the capability for new services is added at a later date. This flexibility should permit incremental growth of the system without making major portions of the prior installation obsolete.
- To incorporate "decision milestones" in the franchise that interrelate technological and economic feasibility with the introduction of new services and also the control of those services. These decision milestones may govern such considerations as (a) under what conditions new services are introduced, and how the costs are apportioned; and (b) whether some services that cannot be sufficiently defined when the franchise is granted are to be treated as the equivalent of separate franchise or licensing negotia-

tions in the future. In this category are many public and institutionally oriented communication services.

Many tradeoff considerations will complicate the application of this approach, and the complexities increase with the size of the contemplated cable system. The basic intent, however, is to reduce technological risk when the system is later expanded, by judiciously structuring not only the cable system design but, more important, the franchisor-franchisee relationship and their respective responsibilities. In effect, it is possible—and wise—to defer decisions that cannot be made sensibly at the time of the initial award of the franchise. There will be some penalty, but it should be less severe than the penalties inflicted by premature and inadequately justified decisions.

II. AREAS OF TECHNOLOGICAL UNCERTAINTY

BACKGROUND

Before the franchising approach proposed herein can be evaluated, it is necessary to have at least a rudimentary familiarity with some of the areas of technological uncertainty applicable to the current cable TV industry. Each such area will affect the structuring of the franchise.

Even without being able to forecast which new nonbroadcast services will develop first, one can be sure that cable's broadband communications capability (making available many spare channels) will open new vistas for cities. Basically, the coaxial cable resembles a telephone in that both transmit information in electrical form, but the cable far outstrips the telephone line in communication capacity. Because TV signals require a channel capable of carrying a large volume of information, the cable is inherently a broadband device (high volume of information per unit of time). One cable TV channel, for example, can transmit as much information in the same length of time as 1500 to 2000 voice-grade telephone lines. What is equally significant is that since cable transmission does not produce external radiation, no FCC frequency spectrum allocation is required. To add channel capacity it is necessary only to add more cables (and associated equipment).

Thus, the coaxial cable is the information equivalent of a broad highway. If the highway is paid for largely by subscriber charges for commercial TV programs, which use only a few "lanes," the remaining lanes are available at a relatively low incremental cost.[1]

Broadcast radio and television stations have only one channel each and are usually forced by economic factors to program for the largest possible audience, neglecting minority or specialized interests and needs. The cable's spare-channel capacity, which converts the present "economics of scarcity" to an "economics of abundance," is the primary focus of future interest. For the cities, such capacity represents a unique opportunity to achieve a quantum improvement in urban telecommunications at a reasonable cost.

The attraction of obtaining almost a "free ride" on the spare channels of the cable system causes much of the franchising difficulty because of the tendency to

---

[1] This presumes that the extra lanes are used for communication services that are compatible with the basic system design, components, and reliability. Such is often not the case for many desired future services.

confuse the physical construction of the system with the jurisdictional responsibilities, which are largely determined by how the system is used and who pays the cost.

Thus the 20-channel cable system specified by the FCC ostensibly offers many communities ample capability for new services. The offer is qualified in many ways, however.

It already has been noted that the new services may require different kinds of equipment and changes in cable network configuration. As another example, future services may require a much higher quality system. The CATV industry grew by supplying a reasonable TV picture to customers who could not receive one directly or who received one that was grossly unsatisfactory. It has been possible to do this with relatively unsophisticated hardware that was not noted for its reliability. The standards applied were generally those of the consumer TV equipment industry (stressing low cost and high-volume production) rather than the more stringent ones of the telephone industry, where reducing the need for service gets extremely high priority because the same company supplies the network, the equipment, and the services on a continuing basis.

Despite publicity, catalog sheets, and brochures that suggest the contrary, much of the cable system equipment necessary for new applications exists only in developmental or prototype form. (All of the systems now under construction for two-way digital communication services, for example, use different, and in some cases custom, components.) Furthermore, the reliability requirements, if any, are an internal matter for each manufacturer, since no external standards have been established. For the few major-market cable systems in operation, the incidence of service complaints appears to be substantial.

Reliability, though always desirable, is perhaps less critical in TV program retransmission; but for a new service, such as a community burglar alarm or even a meter-reading system, it must be a prerequisite to implementation.

Thus, franchising authorities must be aware that many services may be precluded if the initial system design is not reliable enough. A "paper" cable system design supplied as part of a franchise application may indicate future expansion to achieve any services the community might desire, but in fact such expansion could be virtually useless. For example, many cable systems use power supplies that can introduce or couple short, transient pulses of interference. For television viewing this may not be too objectionable, but the pulses could cause numerous false alarms in a sensor-activated fire-alarm service. In short, a reliability level adequate today may be unacceptable in the future. Because of such factors, many systems for which franchises already have been granted, or are now in the process of being granted, will fall far short of their technological potential.

Given the dynamic state of cable technology and the present ambiguities of regulation, it is probably not possible to avoid *all* major problems. One simple option, therefore, is for a city or county to hold off granting a franchise until the situation stabilizes. Indeed, many consortia, formed either on a regional basis (city-county or among a number of cities) or a functional basis (city government, school district, community agencies), are already in existence, with their major objective being to delay precipitous action in granting franchises until mutual agreement is reached on the kind of cable system desired.

Such investigation and review is useful in establishing *initial* system requirements. However, because there is a gap between technological and economic feasibility, and an acute shortage of "hard" data that can relate the two, there is a limit at present as to how far the definition of *future* requirements can go. Consequently, at some point, the decision must be made to delay granting the franchise or to proceed upon the best information available, even though admittedly inadequate.

Obviously, the preferred choice would be to proceed, if it appears possible to minimize the inherent risks and to establish the technical, functional, and regulatory requirements of the cable system in a manner permitting system growth and modification to be both responsive to community needs and reasonable in cost. To achieve this in the franchise is no small task. It is similar to developing a major aerospace program plan and set of specifications in which an interdisciplinary knowledge of technical, economic, and requirement factors is necessary.

For the cable system, it is likewise necessary that both the potential advantages and the limitations of cable technology be appreciated by representatives of the franchising city (and interested community organizations) so that realistic decisions can be made.

The following portions of this section present the highlights of current areas of technological uncertainty in order to relate them to the process of structuring a franchise. The presentation is by no means exhaustive; it should be supplemented by referenced or other bibliographic material. In many cases, competent consultant assistance may be advisable.

TECHNICAL STANDARDS

To understand the extent to which technical standards are currently established for cable TV systems, it is useful to review briefly the FCC standards that became effective on March 31, 1972.

Four classes of cable TV channels are defined (see Appendix A for FCC definitions). Class I includes those that redistribute commercially broadcast TV programs. Class II covers cablecast programs, i.e., those that originate within the cable system. In this category at present are the local origination, public access, education, and government channels. Also included would be leased channels, but not those that employ decoders or other devices for restrictive viewing.

Class III includes all new services and forms of communication for which the cable system might be used in the future, but is limited to one-way information transfer from the cable-system distribution center to subscribers. This covers coded or scrambled signals, facsimile, and many as yet undefined services.

Class IV applies to "return," "response," or "upstream" communications, in which the information, in whatever form, flows from the subscriber to the central distribution center of the cable system. Thus, the return portion of all two-way communication services falls into this category.

So far, the FCC has established technical standards for only Class I channels (see Appendix B). The standards cover some of the signal characteristics necessary to provide each subscriber's TV receiver with a picture and sound whose quality approximates that received off the air in an average reception area. In effect, these standards partially insure that the cable system itself will not excessively degrade the original broadcast picture quality. Some degradation may, in fact, occur in cable systems meeting all FCC requirements.[2]

The FCC's purpose in not establishing standards for Class II, III, and IV channels is to encourage experimentation and participation in developing new services and uses for the cable system. This objective is laudable, but the approach does present problems. As an example, the public-access channel (Class II) is currently in use in the two large New York City cable systems, and presumably will be used similarly in all other major systems. If all programs for the participants were prepared at the cable system studios, the technical quality would be adequate, but many programs are being videotaped in the field on a variety of video recorders, many of which are incompatible with each other. This incompatibility can be rectified at the studio by using appropriate editing and resynchronizing equipment, but there is no requirement that this be done. The result may be poor quality or even unintelligible pictures delivered to the subscribers.

Therefore, though the FCC has not yet established standards, the franchising authority in many cases may be well advised to set its own, even if later preempted, to avoid poor performance and resultant complaints.

Over and above the question of establishing the standards and incorporating them into the franchise, is the problem of monitoring performance and assuring compliance with the standards. Many past franchises have included *no* monitoring and compliance requirements. Others have included requirements that turned out to be both ambiguous and unenforceable in actual operation. The FCC performance tests, specified in Appendix B, apply only to major market systems, and they contain no penalty provisions.

Thus, any new franchising approach should include an appropriate set of monitoring and compliance procedures together with adequate technical standards. This is true not only for the initial cable system configuration, but also for future expansion. In the latter case, if the standards and procedures cannot now be determined, the franchise structure should facilitate their inclusion at the appropriate time.

To indicate what this means, Appendix C provides two illustrative provisions taken from actual franchises. The first is a clause from New York City's franchise with TelePrompTer, which represents the total of the technical standards imposed. The requirements are both subjective and incapable of measurement. The responsibility for establishing and maintaining high standards in a case like this rests with the cable system operator. (The FCC standards will not be applicable for five years in this situation, since they were formulated after the New York City franchise and provide "grandfathering" exemptions.)

Not only is the overall technical performance of the system not subject to quantitative requirements, but there may be considerable variation in performance from one subscriber to another, with some receiving substantially poorer pictures. There is no clear way of forcing a correction of such conditions. The city, by weight of its franchising authority, obviously can exert considerable pressure upon the cable-system operator to correct deficiencies. When some threshold level of complaints is reached, it can negotiate improvements. This, however, does not change the fact that both the deficiencies and the corrective actions are not defined.

---

[2] As an example, the standard of 36 decibels for the ratio of visual signal level to system noise and undesired co-channel TV signals may result in visible picture interference in major-market systems. There are no standards as yet for color and sound parameters.

Even where attempts have been made in the franchise to deal with the problem of compliance, they have not been detailed sufficiently. The second portion of Appendix C is taken from a franchise granted in mid-1972 in Redlands, California. It contains specific operating standards and also a "proof of performance" specification that requires testing to demonstrate that the system meets the standards. This at least permits a determination of fact (i.e., the system either does or does not meet the standards) in a relatively straightforward manner. What is completely omitted, however, is the question of what happens if standards are not met, or as is more usually the case, are met only partially.

The only penalty provisions in many franchises are those that call for the cancellation of the franchise. This ultimate sanction might be considered if the system were completely unacceptable, but is no help at all, for example, in a case where perhaps 80 percent of the requirements are met and 20 percent are not. Rather, a franchise should require a range of compliance and penalty provisions to match the range of anticipated test results, with the severity of the provisions being proportional to the severity of the deficiency.

Table 1 shows how this might be done (in oversimplified form). Assume that the performance specifications must be tested at a large enough sample of subscribers' home TV receivers (perhaps 100 to 200), scattered at all locations throughout the system. From these tests, a statistical basis can be determined for ascertaining what percentage of subscribers are not receiving the specified quality of service.

Some tolerance for reasonable operational variations is allowed. As shown in Table 1, if up to 5 percent[3] of subscribers' terminals are not meeting specified performance, then no corrective action would be required, except voluntarily by the system operator.

Above that threshold, however, specific requirements are imposed by the franchise. A period of time is allowed for the operator to correct the deficiency, which, in effect, reduces the number of out-of-specification subscriber locations.

If the operator is either unable or unwilling to comply, then a penalty in the form of a rate reduction (or, alternatively, a fine) is imposed. Thus, the subscribers pay less for degraded service. Concurrently, the system operator is permitted to continue his business but is compensated only in proportion to the quality of service rendered. There is a financial incentive for him to improve the quality, and the relatively rare situations warranting franchise cancellation are reserved for those systems where performance is poor enough to merit this expensive and time-consuming solution.

Obviously, it is difficult to extrapolate such performance monitoring and compliance provisions for future cable services not yet defined. It is reasonable to attempt some provisions, however. As one example, the franchise can require that as new FCC technical standards are issued, they automatically be incorporated in provisions such as the ones in Table 1. A corollary is to write the franchise so that additional compliance provisions are negotiated with each franchise milestone (i.e., with each new service or system expansion approved in the future).

The magnitude of effort necessary to monitor a cable television system for

---

[3] The threshold of 5 percent is purely illustrative. If a particular technical standard is critical, 1 percent or 2 percent might be more appropriate. Conversely, the threshold can be broadened for less important features.

Table 1

EXAMPLE OF FRANCHISE COMPLIANCE AND PENALTY PROVISIONS

% of Terminals Not Meeting Specifications	Days to Implement Corrections	% of Substandard Service Allowed	Rate Reduction Penalty for Noncompliance (%)
0-5	---	---	---
5-20	60	0-5	20
20-30	90	0-5	30
30-40	120	0-5	40
Above 40[a]	---	---	---

[a]Consider franchise cancellation.

compliance varies in proportion to the size of the system and to the different kinds of services rendered. At the low end of the scale would be a small system that would only be used to distribute broadcast TV programs to perhaps less than 3500 subscribers.[4] Here the performance tests might simply consist of testing the signal characteristics at some specified number of subscribers' TV sets, possibly 10. The tests could be completed in one day, and might be performed by technicians of the cable system while being witnessed either by a city employee (e.g., City Electrician) trained for the purpose, or by an outside consultant.

At the upper end of the scale, the testing, monitoring, and compliance requirements for a large cable system become more complex and elaborate. Many more points in the system should be tested, since there may be great variation in signal quality along the lengths of the trunk and feeder cables. Furthermore, the tests that demonstrate adequate quality for broadcast TV programs may not be sufficient for two-way services.

One method of monitoring a major system requires a staff of trained technicians supervised by competent engineers. If, on the other hand, the franchising authority relies on the system operator's staff, then validations must be provided, either by witnessing the performance of the tests or spot checking. In either event, the in-house personnel required by a large city or county may be substantial. At some point, establishment of an Office of Telecommunications, as New York City has done, may be warranted. This issue is discussed in Sec. IV.

The cost for such monitoring is substantial and in many cases can easily exceed the franchise fees that are paid to the city or county. For example, a community of 60,000 people might have 20,000 residences. If 50 percent penetration is assumed, there would be 10,000 subscribers, and at $60 per year, the total annual subscriber revenue of the system would be $600,000. Using the FCC-recommended rate of 3 percent, the annual franchise fees would be $18,000.

Therefore, if even one full-time technician is required to monitor the cable system, the cost of his salary plus overhead support would require most, if not all, of this $18,000. Far from being the revenue "bonanza" that some communities

[4] The cutoff point above which the FCC requires the operator to provide local origination cablecasting.

expect, it is a sad fact that most cities and counties (particularly in the larger markets, where cable systems are more complex) may find themselves compelled to spend more in monitoring and compliance than they receive in franchise fees. This expense could be justified, of course, if the cable system provided enough compensation through public service to the community.

CHANNEL CAPACITY

Since CATV began its operation by retransmitting commercially broadcast TV programs via cable to subscribers, the basic measurement of system capacity has been the number of TV programs distributed simultaneously. Furthermore, a CATV "channel" was equivalent to a standard broadcast TV channel, both in terms of carrier frequencies and information-handling capability. The former corresponded to the FCC-assigned broadcasting frequencies for each TV station, and the latter to the standard TV channel bandwidth of 6 MHz (6,000,000 cycles per second).

Consequently, a cable channel is also 6 MHz wide, and the total channel capacity of a cable system is determined by how many such 6 MHz channels can be accommodated through the system. Even if a particular service of the cable system does not utilize a TV format (e.g., a fire-alarm system), and therefore may not require the same bandwidth as a TV program, the conventional measure of capacity remains the number of equivalent TV channels. (A polling system for 10,000 homes, for example, could be accommodated in a band about 1 MHz wide, or approximately one-sixth of a TV channel.)

Current coaxial cables are capable of carrying a very wide band of signals, although higher frequencies undergo greater loss as they proceed through the cable. The major limitation, however, is the amplifiers needed at periodic intervals along the cable to compensate for the signal losses. Amplifiers presently in production in the United States for cable-system use place an upper limit of about 300 MHz on signals that can be transmitted effectively. This is the equivalent of 50 TV channels, but a practical limit per cable would be 25 to 35 channels due to a variety of signal interference problems [2].

Thus, a cable "highway" can currently transmit information that is the equivalent of up to 35 TV channels. In actual operating cable systems to date, however, this capacity has not yet been reached, primarily because of lack of demand.

Historically, CATV systems originally provided 3 or 5 TV channels, later grew to 7, and then to 12 in most currently operating installations. Each of these increments is tied to some parameter of the broadcast TV industry:

- *3 channels:* Usually one channel for each of the 3 TV network affiliates in the local geographic area.
- *5 channels:* Usually, three network affiliates plus up to two independent stations.
- *7 channels:* The maximum number of VHF stations the FCC assigns to any area. There are a total of 12 VHF channels (channels 2 to 13 on the TV receiver tuning dial), but if adjacent channels broadcast in the same area, there may be interchannel interference. Consequently, the maximum

assignable channels would be numbers 2, 4, 5, 7, 9, 11, and 13. (A frequency gap between channels 4 and 5 permits this exception to the rule.)

12 channels: The maximum number of VHF tuning positions on the TV receiver. Even though only 7 can be receiving off-the-air TV broadcast programs, recent cable systems have used the other 5 positions by bringing in the UHF or nonlocal VHF programs (converting them to an appropriate channel frequency) or have provided cablecasting in order to utilize these 5 channels.

The 12-channel cable system represents a natural plateau if standard TV receivers are used as the display device. It is important to realize that this is due more to FCC broadcast assignments, which relate to the available frequency spectrum for radiated signals, than to an inherent limitation by the cable component, which, as noted above, can currently carry 2 to 3 times this capacity.

TV receivers built in the last few years also include UHF tuners for channels 14 to 84. Since UHF frequencies (470 to 890 MHz) are too high for present U.S. cable systems to carry directly,[5] the UHF tuner has not been utilized for cable TV applications.

The 12-channel VHF dial has become, therefore, a standard channel-selection device for all cable-system transmission compatible with broadcast TV standards. For cable systems offering a capacity of more than 12 channels, a variety of switching and conversion techniques have been developed to work in conjunction with the TV receiver VHF tuner.

Before reviewing these, however, one problem should be noted that afflicts cable systems with both less and more than 12 channels. This is termed "direct signal pickup" or "on-channel" interference.

During the early CATV days, when cable systems existed in rural areas where the radiated signal strength from TV broadcast stations was very weak, the cable brought the TV program to each subscriber on the same channel position (carrier frequency) as broadcast. Thus, if the CATV picked up Channel 4 off the air, it was delivered as Channel 4.

In the more densely populated metropolitan areas, the radiated signal strength is usually strong for those TV broadcast stations located in the area. Suppose, for example, that a cable system picks Channel 4 off the air and delivers it to each subscriber's TV receiver. Now, however, each receiver will also pick up some Channel 4 signal from the radiated broadcast. This is true even with the receiver antenna disconnected, since a few inches of unshielded wire will pick up the signal in sufficient strength to allow it to proceed through the receiver.

The receiver therefore carries two signals on the same channel, one delivered through the cable and one picked off the air. Unfortunately, these arrive at slightly different times since the propagation velocity through the coaxial cable is less than in air. As a result, the relative signal delay results in a "ghost image," which can be very distracting (Fig. 1).

Once both signals enter the receiver, there is no practical method of separation since they are identical in all essential characteristics. Better methods of shielding

---

[5] Short-length cable systems, such as an apartment house master antenna installation or an on-campus CCTV system, are an exception, since the losses over the shorter cable distances are tolerable. Also, some European systems are designed to distribute UHF frequencies directly, and do utilize the UHF tuner.

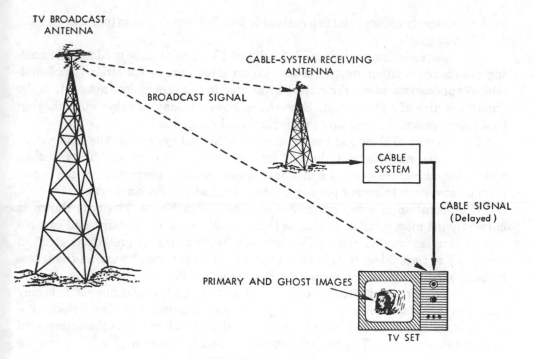

TV BROADCAST ANTENNA

CABLE-SYSTEM RECEIVING ANTENNA

BROADCAST SIGNAL

CABLE SYSTEM

CABLE SIGNAL (Delayed)

PRIMARY AND GHOST IMAGES

TV SET

Fig. 1—Direct signal pick-up interference

at the input to the TV receiver can reduce the off-the-air pickup and in some franchised cities the system operator has done this, but it does not represent an effective solution. If any TV set malfunctions later, the question of whether this modification caused the malfunction can be a source of dispute.

The magnitude of this problem becomes critical for the major-market areas. In Los Angeles, for example, there are 7 assigned VHF stations: Channels 2, 4, 5, 7, 9, 11, and 13. A cable system in the Los Angeles region might well encounter on-channel interference for all seven, which would mean, in effect, that they could not be cabled directly to subscribers at their broadcast frequencies. The FCC requires, however, that all local TV broadcasts be provided to cable subscribers.

A commonly accepted solution has been to translate the carrier frequency of each channel subject to direct interference to a frequency unused in that local area. Channel 2 could be translated to Channel 3, for example, because if Channel 2 were assigned, Channel 3 would not be. This translation can be accomplished at the cable-system headend, and the signal delivered as if it were originally Channel 3.

This works for a few channels, but in the Los Angeles case noted above, it is not possible to do this for all 7 channels, since there would only be 5 unused positions left on the TV tuner dial. The situation would be even further aggravated by the requirement for cable-casting channels (government, education, etc.), which could also not directly use the seven positions where strong off-the-air pickup existed.

Consequently, cable systems that do not utilize converters at the subscriber's location ("set-top converters") are limited to less than the available 12 channels by the direct pickup interference problem. For the large cities, a nominal 12-channel system can easily reduce itself to a realizable 5-to-7-channel system (per trunk cable) for Class I signals.

(The channels that are not usable for direct TV signals can be used for cablecasting automated weather reports, stock market data, etc., or for other "black-and-white" applications where the picture presented to the viewer does not contain the "tone" features of a photograph. Even these, however, may be subject to annoying beat interference, showing up as periodic visual distortion.)

The current FCC regulations require a 20-channel system for the major markets, with one nonbroadcast channel for each broadcast channel.[6] Thus, such a system would require either a multicable system with no converters, or the use of set-top converters to permit greater channel utilization of a single cable.

The operation of a set-top converter is shown in Fig. 2. The type shown is currently the most popular, costing as little as $25 to $30 in quantity, and capable of receiving between 25 and 35 TV channels. Many channels are "translated" in frequency at the cable-system headend so that all frequencies lie within the transmission band of the cable system and there is no duplication of frequencies.

A tuner is provided with the converter to permit selection of channels. This may either be the familiar rotary type or a group of push-button switches. Selection of any channel tunes the converter to that incoming signal, which is then converted to an unused standard TV channel frequency (usually Channel 12 or 13, since one or the other will be unassigned in any geographic area).

This converted signal then enters the home TV set as if it were Channel 12 or 13, and the receiver, when tuned to that position, will display the picture. In similar fashion, any other selected channel is also converted to Channel 12 or 13. Thus, the TV tuner is left fixed and all channel tuning is performed at the converter.[7]

With this technique, there is no longer any problem of on-channel interference, since the only signal reaching the TV receiver from the cable system is a channel that is not locally broadcast.

The use of the converter, while eliminating on-channel interference, unfortunately introduces new and different possibilities for interference and picture degradation. Many converters respond inadequately to variations in signal strength. A strong signal can "overload" them and cause picture distortion. Harmonic frequencies from the low-band channels can affect the higher frequency channels, and spurious oscillation from both the TV receiver and converter may also distort the delivered picture. The use of components designed to minimize this interference (e.g., push-pull amplifiers) is advisable, but does not offer a complete solution.

Over the past few years, converters have been responsible for a large proportion of customer complaints. So-called "second-generation" converters with considerably improved performance in some areas (but degraded in others, such as frequency drift) have been introduced in the last year. Converter designs are still aimed at the lowest possible cost since, even at $25, they may represent 20 percent of the system

[6] The additional requirement to carry "mandatory" broadcast signals, could, in certain regions, result in more than 10 broadcast channels and, consequently, more than 20 channels for the system.

[7] If the TV set itself incorporated a 25 to 35 position tuner rather than 12, the set-top converter would no longer be needed. Although such "CATV-compatible" receivers can be developed easily, the market has been too small until recently to attract the major TV receiver manufacturers. Also, many standardization and compatibility problems, particularly in channel frequency assignment, must be overcome. Nevertheless, one manufacturer, Magnavox, has recently announced its intent to start production of a 31-channel "CATV-compatible" TV set.

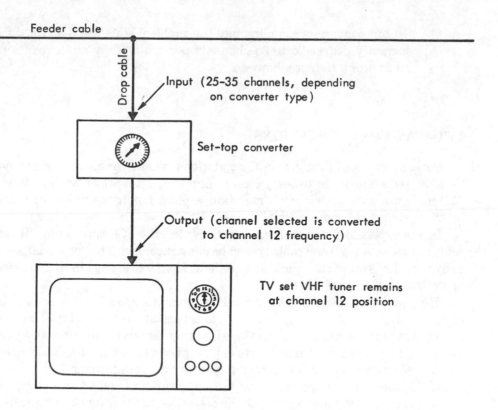

Feeder cable

Drop cable

Input (25-35 channels, depending on converter type)

Set-top converter

Output (channel selected is converted to channel 12 frequency)

TV set VHF tuner remains at channel 12 position

Fig. 2—Operation of tuneable set-top converter

cost per subscriber. Because of the low-cost emphasis, some deficiencies are still tolerated, and there remains a high potential for malfunction. Converter performance and (most critically) reliability, must be carefully specified.[8]

In summary, the key points affecting channel capacity are the following:

1. Current U.S. technology utilizes coaxial cable systems that can transmit 25 to 35 channels per cable at VHF frequencies up to about 300 MHz.
2. TV sets can directly select one of a maximum of 12 VHF channels. This establishes the upper limit per cable, if no converters are used.
3. This 12-channel limit must be reduced by the number of strong broadcast stations in the area, since simultaneous reception from the cable and off the air will produce unsatisfactory pictures due to ghost images. In the worst case, the 12-channel maximum may decrease to only 5 channels usable for normal TV signals.

[8] There is another group of cable configurations, termed "switched systems," in which set-top converters are not necessary because switching and selection of any desired channel are performed at a distribution center rather than in the subscriber's home. Although switched systems have some advantages for certain applications, their major disadvantage is the complexity of cabling and the necessity for local switching centers, which pose a problem for urban areas. They have been used in the United States only in a few pilot installations and are not a significant part of current cable operations.

4. Set-top converters can eliminate the problem stated in 3 above, and can currently deliver 25 to 35 channels per cable. They do introduce other kinds of interference, however.

TWO-WAY COMMUNICATIONS

Perhaps no area of the new FCC regulations is more confusing and ambiguous than the requirement for two-way communications. The specific wording (Par. 76.-251) is, "Each such system shall maintain a plant having technical capacity for nonvoice return communications."

In the explicatory section preceding the rules, the FCC comments, "It will be sufficient for now that each cable system be constructed with the potential of eventually providing return communication without having to engage in time-consuming and costly system rebuilding."

The opinion of the FCC is that it would be premature to specify the exact nature of the two-way communications capability, for at least two reasons. First, the market for two-way services has not yet developed, so that requirements cannot be defined; and second, the technology is in a state of flux, preventing standardization of either techniques or components for achieving two-way communications.

While these points are valid, the present wording leaves the method of implementation almost completely up to the discretion of the franchisee, regardless of the extent of future two-way services contemplated by the franchising authority. For example, the letter of the FCC rules can be satisfied by connecting to each subscriber's home a simple wire that can transmit only an "On/Off" or "Yes/No" signal.[9] This may permit audience polling, but is useless, for example, for interactive educational television in which students watching an instructional program are supposed to make comments and ask questions.

In such a case, the franchisee may rightly claim that the system can be modified later to provide a greater range of two-way capability as each new application evolves. The vital question, however, is at what cost and with what degree of flexibility such modification can occur. Unless this question is explored in detail at the time of the initial franchise, many future options may be unwittingly foreclosed.

To understand the implications, a brief review of current two-way cable technology is useful [3]. At present, two major approaches are possible: (1) furnish a separate cable for reverse transmissions; or (2) attempt to send signals in both directions simultaneously over the same cable, using different frequency bands or time-division techniques to separate the "downstream" messages (headend to subscribers) and "upstream" messages (from any subscriber terminal to the headend).

The highest-confidence-level approach is to provide a separate cable (or cables) for upstream transmission, since this entails fewer technical problems. The initial system cost will be higher, since the upstream cable will be installed at the same time as the downstream cable. If the need for upstream communications is not

---

[9] Many operators are installing an extra ("shadow") trunk cable for future two-way use, but with no feeders, amplifiers, etc. Until these components are provided in such cases, the system has no operational two-way capability.

evident at an early date, upstream amplifiers could be installed later, when required.

There may or may not be some wasted cable bandwidth capacity with this approach. At this point in time, it is difficult to visualize using 30 to 35 channels for reverse communications. A security surveillance system, for example, might be an application in which several TV channels would be required if cameras at different locations transmitted pictures to a common point. Institutional communication services such as school-to-school, hospital-to-clinic, or business-to-business communications may eventually require substantial upstream bandwidth. For services oriented primarily to home subscribers, however, there may be no need for a wide-band upstream capability. Audio or digital messages requiring a relatively narrow bandwidth will probably be adequate. For example, a school can deliver an instructional TV program to at-home students, but permit only audio queries from students to the originating classroom or studio. Polling-response services, as a second example, require only digital messages in each direction.

The separate cable approach, therefore, carries a higher initial cost,[10] and may provide more reverse communications capacity than is needed in the near future. Its great advantage, however, is that it will work with more certainty than the alternative.

The alternative technique of sending signals both ways over the same cable has not as yet been proved technically satisfactory in large-scale, operational cable systems, and carries a greater technical risk.

The method that has been used most frequently is illustrated in Fig. 3. The downsteam and upstream signals are run through the same cable, but at different frequencies. The cable is bidirectional and presents no problems, except that the amplifiers in the cable system are unidirectional devices. Consequently, a "bypass" path is provided using filters to permit reverse transmission around each amplifier.[11]

Since the same cable is used, a low-cost initial installation can be achieved, and for this reason the approach is highly popular with the cable-system operators. There are two strong disadvantages, however, that should be known by the franchising agency.

[10] The initial cost would perhaps include 10 to 20 percent more for the extra cable and installation labor. The added cost for amplifiers for the upstream cable would be on the order of $2000 per mile, partially compensated by the fact that the downstream cable amplifiers would cost less without the bidirectional capability. There would also be much less labor expended in "tailoring" the system, now necessary for single-cable bidirectional systems.

[11] As shown in Fig. 3, a complementary pair of "cross-over" filters is placed both in front of and behind each amplifier. Each filter accepts all input frequencies and has both a high-band and a low-band output. The high-band output is connected to the downstream amplifier input, while the low-band is connected into the reverse amplifier (shown in dashed lines).

Normally, the high-band would start at 54 MHz (the lower end of Channel 2) and include all higher frequencies, so that all broadcast TV channels could be carried. This restricts the reverse, low-band spectrum to frequencies below 54 MHz. In practice, a bandwidth of perhaps 5 to 30 MHz is available for upstream communications, the equivalent of 1 to 4 TV channels.

The approach (termed "subsplit" or "low-split") has strong economic appeal since it uses a presently unused portion of the cable bandwidth (below Channel 2) for reverse communications, and also allows the upstream components to be installed at some later date. Thus, room can be left physically for the dashed components of Fig. 3, and theoretically they would be added when required. Indeed, some currently available two-way amplifier housings provide space to add modules for this purpose.

Almost all pilot two-way cable systems constructed recently or now under construction have used the low-split approach for polling-response services.

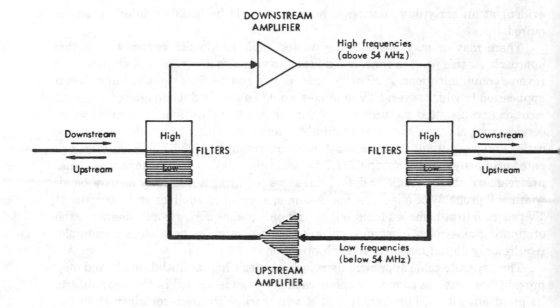

Fig. 3—Simultaneous two-way communication on a single cable

First, reliable and interference-free operation remains to be demonstrated in large cable systems involving a number of cascaded amplifiers. Pilot experiments (using the frequencies below 54 MHz for reverse transmission) have indicated that the upstream signals caused distortion of the normal downstream TV pictures. (This is less evident when only a narrow-band digital signal is being sent upstream rather than a wider-band TV signal.)

One problem is that the required filters present extremely difficult design problems. Even small variations in filter characteristics, when multiplied by the number of amplifiers along the length of the trunk cable (since a filter pair is required at each amplifier location), can add up to unacceptable picture distortion.[12] While these difficulties can be expected to be minimized in the future, the technological risk remains considerable, and system designs that incorporate this approach should be reviewed carefully.

Second, the "low-split" approach limits upstream communications to a total of 3 or 4 channels. While this is more than ample for any applications that might be implemented in the short-term (primarily digital-polling-response subscriber services), it is by no means evident that such capacity will be adequate over the life of the cable system franchise. Even a single security surveillance system might well require 3 to 4 video channels so that pictures can originate from several locations. Such an application would, by itself, use up almost all of the available upstream capacity.

Conversely, if there are no plans to utilize wide-band two-way services for an indeterminate number of years, it may prove economically attractive to retrofit the

---

[12] Other potential sources of picture degradation include coupling through common elements such as power supplies, and interference introduced by automatic-gain-control circuitry.

system later, even if such retrofit involves considerable extra cost. The "real" cost of providing major capital facilities that lie unused for many years must take into account how that money could be used for other beneficial services.

The tradeoff will depend on the status of planning for two-way services, both by the system operator and governmental agencies. If a preliminary survey indicates high interest, the system design should provide a more adequate margin of upstream transmission capability.

Recently, many of the newer cable systems have featured two trunk cables, with the second as the shadow trunk. The first cable, with a converter, provides 25 to 35 channels of downstream communications and, using the low-split technique, up to 4 upstream channels. Since this cable alone meets FCC requirements, it is possible to use the second cable primarily for future services, possibly with equal downstream and upstream capacity.

One approach, for example, is the "mid-split" technique, in which frequencies below 108 MHz can be used for reverse transmission on the second cable only. Thus, the shadow trunk (when activated) is in effect split, with approximately 10 to 15 channel capacity in either direction. This technique does provide more upstream channels (probably adequate for foreseeable applications), but again extreme care is necessary to minimize the interference effects of two-way transmission on the same cable.[13]

Thus, in the current state of the art for two-way cable communications, the following points are important:

- Designing the cable system for two-way communications on a single cable includes (a) a substantial technical risk, and (b) the possibility of inadequate capacity for future applications requiring wide-band reverse transmissions. It does, however, represent the lowest initial system cost.
- Providing a separate cable for reverse transmission (a) increases the initial installation cost but (b) reduces technical risk and also offers the greatest capacity for future growth.
- The hybrid shadow-trunk approach of providing two cables initially but connecting up only one for downstream transmission (and, possibly low-split upstream transmission), while reserving the second for future use, is an attempt at a compromise solution. In comparison with a single cable, it (a) offers much more upstream capacity with the "mid-split" technique, but (b) does not solve the problem of distortion arising from simultaneous two-way transmission on the same cable.

SYSTEM QUALITY AND RELIABILITY

It has been pointed out that the transition from a classical CATV cable system, carrying commercial TV programs, to a multiservice communications network may require not only a variety of new equipment but also a much higher-quality system.

---

[13] The more points along the cable system that signals are introduced, the more possibility there is for introduction of noise and interference. This problem is present whether cable transmission is unidirectional or bidirectional, but is more difficult to overcome on bidirectional cables, since it combines with the effects of reverse signals.

"Quality" is a relatively subjective term, difficult to define and measure. Its use in connection with cable technology arises from the fact that CATV system performance has been judged in the past by how "good" a picture was delivered to the subscriber. This is also a subjective reaction, depending largely on a visual comparison with what the viewer is used to seeing on his TV receiver when tuned to off-the-air signals. Quality, however, can be quantified by relating it to specific technical parameters.

Reliability and maintainability are more precise terms. Through experience with telephone equipment and systems, for example, procedures and techniques have evolved by which reliability and maintainability can be quantified, and thus measured and verified (see Appendix D).

The relevance of these terms and measures to a cable television system is now becoming evident for the major markets. Some of the new nonbroadcast services will demand a level of reliability that is not currently being met by the CATV industry. Even a routine utility-meter reading service demands a highly reliable and error-free system, while a fire-alarm system could not operate at all with a high incidence of false alarms or equipment failures. It is therefore of concern to franchising authorities that any system franchised be capable of meeting the higher standards.

Reliability is particularly critical in the early stages because it is extremely difficult to "add" at a later date. The initial system design is the primary governing factor (e.g., choice of components, test-point availability, redundancy, etc.), and if high reliability is essential, it usually must be an inherent original criterion.

Furthermore, there are various aggregation levels of reliability in a cable system. At one end is the malfunction that interrupts service to only one subscriber; at the other extreme is the catastrophic failure of the entire system, which, for example, might happen as a result of electric-power loss at the headend. Each level may warrant individual consideration in terms of specifying reliability and maintainability requirements.

At the present time, it is not easy for a city to determine what degree of reliability or maintainability is necessary for a cable system, since the new services are not defined. It appears important to make the effort, however, for at least two reasons.

First, the reliability of past and current cable television systems seems marginal even for services now being offered. Many systems provided poor-quality pictures even on initial installation. In the few large city systems in operation, subscriber complaints have been as high as 0.5 to 1 percent per day (for a 30,000-subscriber system, 150 to 300 service calls a day). While a large percentage of the calls are undoubtedly not related to cable system malfunction (e.g., the trouble may be in the subscriber's own TV receiver), this must to some extent be indicative of system reliability and maintainability. An essential public service, or even a business-oriented service such as transmitting stock market information to customers, could not maintain continuity of operation with this percentage of service complaints.

Second, if pressure for improved reliability and maintainability does not come from the franchise authorities, it may not come at all or, alternatively, it may vary drastically among systems. (The latter may be a particular irritant for large cities or counties where multiple franchises are granted.) The FCC technical standards offer no assistance, since they merely establish performance characteristics relating to picture quality that must be measured annually and that bear no relation to day-by-day system reliability.

The cable-system operators and the equipment manufacturers desire reliability, of course, but the tradeoff against higher cost is always an inhibiting factor. Unless higher standards are imposed by the franchisor, it will be difficult to stimulate sufficient motivation for increased reliability, particularly for new systems whose costs appear to be mounting rapidly due to other considerations.

Given the foregoing brief introduction to the terminology, we can now discuss some of the factors that influence the quality and reliability of a cable television system. Again, only the highlights can be mentioned; any specific system requires a more intensive study.

Figure 4 illustrates the major components of a typical cable television system designed for one-way information flow. Components from the system antenna through the home TV receiver have been numbered from 1 to 12. Each of these represents a potential source of degradation of the delivered picture, as outlined below.

1. **Antennas.** Although most cable systems locate their antennas to receive the strongest off-the-air signals, it is important to realize that the signals may already be degraded to some extent when received, depending upon the path(s) taken by the signal from the broadcast TV transmitter to the cable-system antenna.

Ideally, the best picture would be provided by a suitable cable link from the TV broadcast studio to the cable-system headend. A few broadcasters have done this voluntarily (in smaller communities) to improve their signal on the cable. Given the adversary position of the two industries, however, this is not likely to happen soon, to any significant extent.

If the antenna cannot be located in a strong signal area, "preamplifiers," which are mounted directly on the antenna tower, are sometimes used to boost the signal strength.

The preamplifiers can introduce significant electrical distortion and "noise" into the system.[14]

2. **Headend.** A large quantity of signal processing equipment is located at the headend facility. Included are components to translate channel frequencies, adjust picture and sound levels, compensate for unequal cable losses, and switch programs into different channels. Both active and passive components are used. There are many sources of noise and potential distortion, and this portion of the system is crucial in its effect on picture quality.

3. **Studio.** The cable-system studio introduces cablecast programs into the system. Assuming that the equipment chosen is adequate, no significant noise or distortion should be introduced at this point. Exceptions may arise from "media translation" equipment converting film or magnetic video tape to electronic form.

---

[14] Electrical noise can be defined as random disturbances in the flow of electrical current, whose effect is typified by "static" in a radio broadcast, or "snow" in a TV picture. All components in the path of current flow will generate some noise, but the most serious are the "active" components (those that require electrical power to perform their desired function such as amplifiers and power supplies.

Coherent or nonrandom disturbances are unwanted signals that arise from manipulations performed upon the electrical information. Amplifiers, for example, do not amplify all portions of a signal equally; consequently, the output signal may be a distorted version of the input, leading to the generic term "distortion."

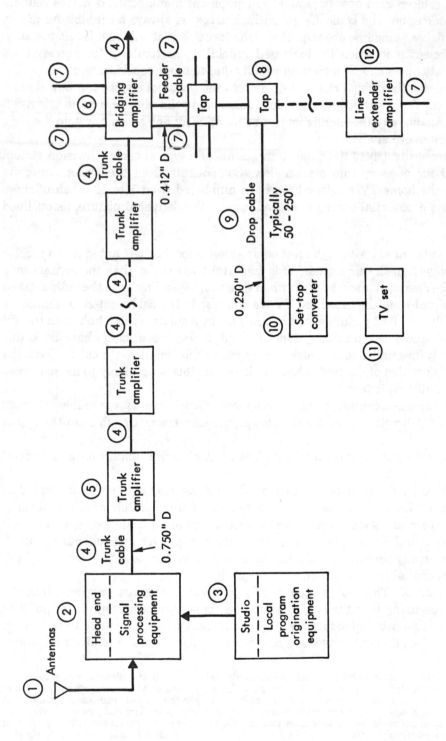

Fig. 4—Sources of signal noise and interference

Film chains inherently introduce some picture degradation, and videotape record/-playback units suffer from a variety of incompatibility problems.

4. Trunk Cable. The coaxial cables used for the main cable-system links are usually 3/4 inch or 1/2 inch in diameter. The larger 3/4-inch type introduces less loss to the flow of information signals, but naturally costs more per foot and is more difficult to install.

All electrical signals suffer some attenuation as they progress through the cable. The loss is greater for higher frequencies;[15] loss also increases with rising temperature.

The attenuation is severe enough to require a number of amplifiers along the trunk cable to boost the signal strength back up to usable levels. (As an example of the magnitude of the loss, a 300-MHz signal proceeding through a 1/2-inch cable will lose half its original power over a 200-ft length.)

5. Trunk Amplifiers. A key design tradeoff is the matching of trunk cable and amplifiers. Using larger-diameter cable reduces attenuation losses, so that amplifiers can be spaced farther apart. The cable cost increases, therefore, while the amplifier cost per mile decreases. Conversely, a smaller cable means lower cable cost but higher amplifier cost.

Apart from balancing costs for any specific system, the delivered picture quality depends heavily on the amplifiers, perhaps more than on any other single component. This is due to the cascading effect of noise and distortion introduced by a string of amplifiers.

The proportion of noise to signal increases with each amplifier in a given "string." Figure 5 illustrates this. The desired signal starts through the cable with a magnitude of X, which reduces to Y at some point down the system because of cable losses. The first amplifier boosts the signal back up to X again and the process is repeated with each successive amplifier.

The same thing happens with the noise content that starts at level A and falls to B. At each amplifier location, however, the amplifier not only brings the noise back up again, but also generates additional noise and adds it to the previous level. Thus, the noise at A1 is a higher percentage of the signal level X than it was at A.

The cumulative process finally results in a noise level that can visibly degrade the picture. It appears usually in the form of "snow." Thus, the number of trunk amplifiers that can be cascaded in one continuous cable run is limited.

The same kind of process also occurs with signal distortion, since each amplifier will introduce some nonlinearity, and the effects are also cumulative.

The specific limit on the number of amplifiers that can be cascaded will depend on amplifier characteristics, cable size, expected temperature variation, and other factors. Typical limits, however, range from 20 to 30 cascaded amplifiers, with spacing between amplifiers of 1500 to 3000 ft.

This means that there is a practical limit to the length of a single trunk cable, which can range perhaps from 5 to 10 miles for conventional trunk cable. Larger diameter "super-trunk" can be run farther. Large-city systems that have considerable distances to cover must be broken up into a number of "hubs," each with its own headend and trunk cable branches.

---

[15] As a comparison, a 300-MHz signal will undergo the same percentage loss in a 200-ft cable that a 50-MHz signal undergoes in 500 ft. Expressed another way, the 50-MHz signal can proceed through 2.5 times as much cable length as the 300-MHz signal before requiring amplification.

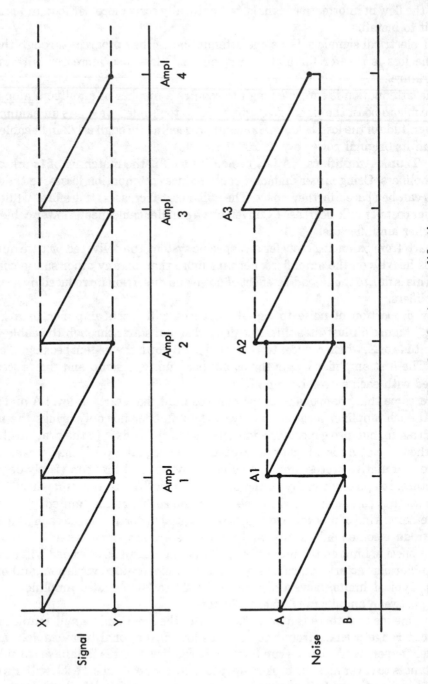

Fig. 5—Signal and noise amplification

6. Bridging Amplifiers. If each subscriber were directly connected into the trunk cable, the large number of discontinuities at the junction points would distort the signals to an intolerable extent. Consequently, it is necessary to electrically isolate the trunk from subscribers' connections. A bridging amplifier is used for this purpose, designed so that any mismatch occurring between it and the subscribers' TV receivers has negligible effect on the trunk cable.

Performance requirements are less stringent for bridging than for trunk amplifiers since they feed shorter lengths of cable, and the amplifiers are less expensive. (Many manufacturers design the bridger units physically as plug-in modules within the trunk amplifier housing.) The possibility of introducing noise and distortion exists in both cases.

7. Feeder Cable. Feeder or "distribution" cable is the term applied to the cable used to distribute signals from the trunk line to the point where each subscriber's individual drop-line connection is tapped off. The cable diameter is smaller than that of trunk cable, usually ranging between 3/8 and 1/2 inch.

8. Taps. Taps are the connecting devices used to couple drop cables and feeder cables. They usually permit up to four drop connections to be made at the tap point.

A tap is a "directional coupler" in that it is designed to offer low resistance to signals flowing from the feeder to the subscriber's TV receiver but high resistance to signals attempting to flow the other way. (In one-way systems, it is necessary to keep the subscriber's receiver or converter oscillation from being injected into the cable party-line.)

9. Drop Cable. The drop cable is the small length of cable used to bring the information directly into the subscriber's home. It is generally about 1/4 inch in diameter. Connection is made to a set-top converter, or if no converter is used, to the TV receiver through a small transformer that matches the characteristics of the drop cable to the input of the TV set.

10. Set-Top Converter. As has been noted, the set-top converter is a critical component in terms of potential interference to cable-system operation. The converter incorporates an internal oscillator (the equivalent of a small transmitter), whose signals are mixed with the incoming channel signals to produce a constant output channel frequency.

Many undesirable effects can occur. Oscillator signals leak back into the cable system or radiate directly into nearby TV sets. Unwanted combinations of frequencies, harmonics, etc., are produced and distort the receiver picture. Oscillator frequencies can drift with temperature or time. To date, a significant proportion of the quality degradation problems noted in cable systems has been due to faulty converter operation. Many of these problems can be alleviated with a higher-cost design.

11. TV Set. The TV receiver is also a major source of difficulty, and one that is outside the cable-system operator's control. Receivers also incorporate local oscillators, similar to converters, presenting the same potential for interference. Their selectivity[16] is poor, and the lack of shielding has been commented upon in discussing the "on-channel" interference problem.

---

[16] This is the ability to tune sharply to a desired channel, so that signals from adjacent channels are completely excluded. Converters, including the newer models, also suffer in many cases from inadequate selectivity.

12. Line-Extender Amplifiers. Line-extender amplifiers are used at the peripheries of a cable system. They allow flexiblity in adjusting the ratio of feeder-to-trunk cable for lowest cost, and they provide a minimum signal level to those remote subscribers at the "end" of the distribution system. Usually, these amplifiers are the cheapest in the system, operating in most cases at lower gains and output levels.

Many of the potentially troublesome components described above become even more critical for two-way communications. A typical example is the cable itself. If the same cable is used for bidirectional signals, the upstream messages may be introduced into the cable at dozens, perhaps hundreds, of points. Poor connectors and faulty splices at many points can cause the cable to act like a huge antenna, funneling noise into a cumulative stream back to the headend and interfering seriously with both upstream and normal downstream transmission.

This means that while a one-way cable installation can operate adequately with "sloppy" design and construction features, a two-way system using the same cable both ways cannot. This is particularly important for franchises that intend to use a cable for downstream transmission only until two-way services are defined at a later time. When the upstream capability is actually required, the system may need an inordinate amount of "tailoring," upgrading, or possibly complete rebuilding, since the quality standards of the initial installation were based primarily on the more tolerant one-way criteria.

Interaction is reduced[17] if a separate cable is provided for upstream transmission. Essentially, then, the upstream cable is also a one-way system with the added complexity of multiple origination points. While this requires care in design, it is simpler than trying to solve both the interaction and the multiple origination point problems on the same cable.

One factor affecting both reliability and cost, which does not show up in Fig. 4, is the question of physical location of, and access to, components. This leads immediately to the major issue of aerial versus underground installation.

From the purely esthetic viewpoint, almost all communities favor underground utilities. While some very highly urbanized areas have actually accomplished undergrounding on a major scale, many still have predominantly aerial telephone and power systems. In some cases, as existing areas are redeveloped, these utilities are relocated underground, but the cost of undergrounding on any massive scale as a separate project is so prohibitive that the conversion process is exceedingly slow.

Cable systems are also dominated by cost considerations. Depending upon the kind of area and the degree of previous undergrounding of utilities (e.g., whether duct space is available), the cost of an underground cable system can range from 3 to as much as 20 times the cost of the equivalent aerial system where the cable equipment is strung on existing telephone or power poles. As an example, in New York City, underground costs of $20,000 to $30,000 per mile have been experienced where duct space already exists, and up to $100,000 per mile where it does not.

This has been an effective barrier to franchise requirements that would compel undergrounding of cable in advance of other utilities. In most cases, the aerial cable is permitted where existing utilities are aerial, and must be underground only where existing utilities are underground.

---

[17] It is not eliminated completely. Components such as power supplies, and common connections such as grounds, can introduce "crosstalk" (interference from one cable to another).

This focus on cost does have penalties with respect to reliability and maintainability. Temperature and other environmental factors are more extreme above ground, and there is greater possibility of accidental or deliberate tampering with the system. Aerial installations may not be practicable at all for certain services such as fire or police alarm systems, since it would be easier to gain unauthorized access to components.

In practical terms, what does the foregoing review mean to a franchisor? The general response is that for the new major-market systems, the questions of quality, reliability, and maintainability must be examined before a franchise is granted if the system is expected to live up to its technological potential. This has not heretofore occurred to any great extent in the franchising process.

Specifically, such examination should include a balancing of costs with respect to new and quantitative reliability requirements. It may well be that certain reliability goals are not worth the added costs, but the franchising city should be aware of the tradeoffs, particularly what type of future capability it might be giving up.

The evaluation should include:

- Specifying reliability and maintainability objectives to be used in the system design process.
- Specifying reliability and maintainability practices and procedures to be used in constructing and operating the system.
- Testing and compliance provisions, ranging from inspection of system installation (similar to the present process of building inspection) to keeping records of, and reporting on equipment malfunction and failure.
- Methods of linking the rate structure or appropriate penalties to the reliability history of the system.

Such an evaluation may well require outside assistance, since many franchising authorities are not staffed for such a task. Costs can be significant. One county with a population of about 250,000 received estimates of 2 to 4 man-months of consultant time to conduct a similar evaluation.

INTERCONNECTION AND NETWORKING

The interconnection and networking of cable systems has not been a major consideration until recently. Rural CATV systems providing only commercial TV programs had little if any need for interconnection.

For large major-market systems furnishing a variety of new services, however, interconnection and networking do become much more important. As one example, a large-city system might be split among several franchise holders (New York City has two in Manhattan alone and may have perhaps a dozen when the other boroughs are franchised). Obviously, there will be some programming, such as the education channel offerings, that should be distributed to all the city's subscribers regardless of franchise area. Thus some means of sending signals from a single origination point through all of the cable systems is necessary.

Even within a single franchise area, interconnection may be desirable. Again taking the New York City franchise as an example, TelePrompTer is required to subdivide its cable system into no less than ten "subdistricts." Interconnection should have the capability either to send the same signals to all subdistricts, or to send different programs to any individual subdistrict or group of subdistricts. True "community-oriented" programming will be permitted by such interconnection and switching capability.

Apart from jurisdictional interconnection requirements, there may be strong financial incentives for interconnecting two or more systems, or for "networking," which is the ability to provide the same program simultaneously to a large number of systems. A Multiple Systems Operator (MSO) may, for example, spend a considerable amount of money preparing a program suitable for the local origination channel. If this program can be used by all of the operator's systems, the economic benefits are very attractive.

The program may be preserved on videotape or film and duplicates physically sent to each cable system. This is feasible, but there are many cases where simultaneous cablecasting is preferable, and some form of electronic interconnection is indicated.

As in any area of telecommunications, the methods of achieving interconnection fall into either of two broad categories, wire or wireless transmission. Interconnection by wire means another coaxial cable link, similar in most essential respects to the trunk cables. The same advantages and restrictions apply. For broadband transmission, where a large number of TV-format channels are carried, it has been pointed out that there are severe limits on the lengths of a cable link; in perhaps 5 to 10 miles, the signal distortion becomes unacceptable.

For one channel (or a few at most), cable links can be much longer, since the channel frequency can be translated to a value low enough to reduce cable losses. In this case, a cable can link points hundreds of miles apart. The coaxial cables that carry commercial TV network programs across the United States are an illustration.

Wireless interconnection, using the public air waves, is governed by FCC allocation of the frequency spectrum, which drastically limits the available options. Commercial microwave links, available from AT&T and other common carriers, offer certain classes of service, but are not designed specifically for cable television system needs, and the charges are relatively high.

The FCC has authorized a special category of microwave link service for cable systems, called "Cable Television Relay Service" or "CARS" (from the former designation of Community Antenna Relay Service). The frequency band 12,700 to 12,950 MHz is allocated to CARS, providing a total bandwidth of 250 MHz.

At the present time, there are three transmission techniques using CARS:

1. **Single-channel FM.** Each TV channel has its own microwave transmitter and receiver pair. The video and audio information is converted to FM form and used to modulate a carrier in the CARS band.

In any one installation, a maximum of 10 channels (requiring 10 transmitter-receiver pairs) can be sent since each channel occupies up to a 25-MHz bandwidth.

2. **Multichannel AM.** For this transmission technique, a block of TV channels is translated directly to the CARS band, and transmitted from a single receiver. The FCC terms this "vestigial sideband AM transmission." The Theta-Com Amplitude Modulation Link (AML) is an example of equipment of this type.

Present FCC allocations permit a maximum of 18 channels (plus 1 "pilot" channel for system calibration) to be transmitted simultaneously via vestigial sideband AM transmission in the CARS band.

3. Multichannel FM. This technique, which has been developed by Laser Link Corporation as their "Air Link" system, multiplexes a block of TV channels into a composite FM subcarrier that is translated up to the CARS band. Again, a single transmitter and receiver are used.

The FCC calls this technique "frequency division multiplexed FM." For sending nine or more TV channels, the entire CARS 250-MHz bandwidth can be used. Theoretically, there is no upper limit to the number of channels than can be so multiplexed, but the quality suffers as the number increases. Laser Link Corporation currently offers equipment for up to 18 channels.

The newer multichannel systems represent a major cost reduction for cable system interconnection. The electronics for an 18-channel link from Theta-Com or Laser Link Corporation is priced in the $85,000 to $100,000 range. For the single-channel approach, one transmitter-receiver combination will cost $7,500 to $15,000, so that an 18-channel system will be 2 to 3 times as expensive as the multichannel equivalent. (Furthermore, 10 single channels are the maximum permitted by the FCC frequency allocations so that an 18-channel system is not possible with this technique.)

Microwave links use antennas that focus the radiated signal into narrow beams, aimed from the transmitter to the receiver. Furthermore, microwaves have characteristics similar to those of light; just as light is refracted or bent as it moves through media of varying density, so microwaves are bent as they move from dense cool air to less dense hot air in an inversion layer. Thus, inversion layers, or fog and rain, can either attenuate the signal or bend it out of alignment to the point where it misses the receiving antenna.

Microwave links therefore have the disadvantage (which cables do not) of being affected by atmospherics, to the point where service may sometimes be completely interrupted. With good design, however, this interruption should be minimal.

In terms of practical transmission distances, single-channel FM CARS links are generally effective up to 20 to 30 miles, and somewhat less reliable (in terms of fading, etc.) at 30 to 40 miles. The newer multichannel systems are lower in power output and shorter in range, perhaps 20 to 30 miles in low-rainfall areas, or 10 to 20 miles in high-rainfall areas.

Microwave also permits interconnection within a cable system when there are physical barriers to economical cable construction. A river, for example, might pose a long detour to the nearest bridge for cable, but would readily lend itself to the wireless link. Another example might be a school that is located far from the main cable trunk lines and that is simpler and cheaper to link by microwave.

Indeed, for the major metropolitan areas, where cabling (particularly underground) is very expensive and where there are many man-made and natural barriers, cable-system construction may prove to be most practical as a hybrid network, including both cable and microwave links.

For interconnection over a large area, regional or national, domestic communication satellites are also expected to play a role. Applications have already been made to the FCC for satellites that could broadcast to "CATV distribution centers," each requiring a receiving facility costing perhaps $100,000. From the distribution center, the programs would be connected, via conventional cable or microwave links, to the network of cable systems associated with that center.

Figure 6 shows one possible version of a community-level educational network. A centralized program origination center is used as the source for a variety of instructional TV programs. These are broadcast via Instructional Television Fixed Service (ITFS), which is low-power broadcast service reserved for educational use, and is allocated the 2500 to 2690 MHz frequency band.

The programs are broadcast via special directional antennas that beam the signal to receiving antennas at each of several cable systems. Here the ITFS frequency is converted to a standard VHF channel and the program distributed to the cable-system subscribers.

Figure 7 illustrates a conceptual regional or national network. Again programs originate at a common center and are linked to a regional receiving center either via satellite, cable, or microwave link. From here they are further linked to individual cable systems.

Networks similar to that in Fig. 6 are now in operation, although on a limited basis, and therefore are of immediate interest in any large city franchise. Regional networks are not now in operation, but may become a reality in perhaps 3 to 5 years.

In summary, the following points are of concern to a franchising authority:

- Major-market systems will require some degree of interconnection. The method of interconnection will affect both the cost and technological capability.
- Regional or national interconnections are not critical requirements currently, but local interconnection is important, both within a large cable system and between adjacent systems that may overlap a jurisdictional entity. (An example is a school district that may not coincide exactly with city boundaries.)
- Selection of the method of interconnection is a tradeoff decision that should be examined as part of the initial system design and franchise structure. No generalized recommendations can be made to cover all cases concerning the choice between cable or microwave links, since the evaluation depends greatly on local geographic and topographic features.
- The cost of interconnection can be a very significant percentage of the total cable system cost.

REQUIREMENTS FOR NEW SERVICES

It has been noted repeatedly that no one can predict with any certainty which new services will prove to be feasible, and among those that are feasible, which will evolve first. This is one major cause of difficulty in structuring a franchise, and a good reason for trying to delay decisions that relate to new services.

It is useful, however, for the franchising agency to be aware of the kinds of capabilities associated with various categories of new services. Even without the desired level of detail, such an association can help in the initial system design and in writing the franchise.

189

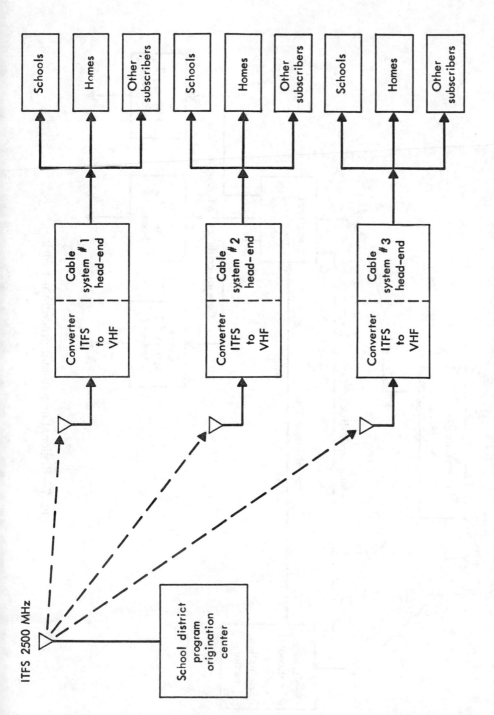

Fig. 6—Community-level educational network

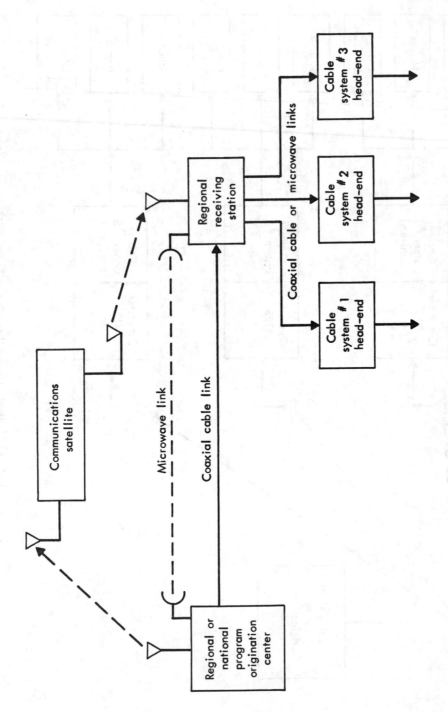

Fig. 7—Regional or national program network

Some services are naturally "compatible" with cable-system configurations. Since most cable systems are designed as large "party lines," a service that requires one central location to communicate with all subscribers, either simultaneously or in sequence, might be expected to be the easiest to implement.

The CATV function of distributing TV signals to all subscribers is a case in point. Also in this category is a group that can be called "subscriber polling-response" services, described in more detail in Sec. III. These are sequentially addressed alphanumeric messages rather than TV programs, but they fit naturally the conventional system design, with the addition of suitable control and terminal equipment.

The reverse form of communications, from any subscriber terminal to a central location, is also compatible with the standard system configuration, although, as noted, it needs more care in design because of the multiple signal origination points. It does not require any major design changes, however.

Sequential, two-way digital communications between the system headend and subscribers can also be performed very economically in terms of bandwidth. A service that can contact perhaps 5,000 to 30,000 subscribers in this fashion might require 4 MHz each for downstream and upstream transmission [4], while a standard 6-MHz TV channel could provide a frame-retrieval service that makes available 216,000 frames of information per hour to 600 separate subscribers.

Transmission of video (moving image) information, on the other hand, uses 6 MHz per channel (if standard TV receivers are the display device). A traffic or security surveillance service, with cameras at many points, could possibly saturate the total upstream bandwidth capacity of the cable system.

Services that require point-to-point communications, from one subscriber to another subscriber, are not inherently compatible with party-line cable systems. They normally would require a switched network, such as the telephone system, which is designed to provide switched paths between designated terminals. To try to provide a service to thousands of subscribers on a conventional cable system is usually difficult.[18] Point-to-point communications among a much smaller number of users, such as perhaps a few dozen schools, hospitals, police precincts, etc., is more practicable. This can be provided on a dual-cable system if sufficient channel capacity is available. Each user could be assigned a separate frequency band and send signals "inbound" to the headend, where they would be routed "outbound" to the reception points. Appropriate converters would permit tuning in the desired messages.

Figure 8 illustrates only a few services that fall into the "compatible, incompatible, and intermediate compatibility" categories. Many other services could also be so grouped.

This does not mean that incompatible services cannot be implemented on a cable system; but that generally such services will be more expensive and awkward to incorporate.

Before a franchise is structured, a possible exercise might be to compile a preliminary "shopping list" of contemplated services from potential users of the cable system, including both governmental agencies and interested community or-

---

[18] One approach might be to send all messages through the headend, but this would restrict contact to small "slices" of the available time.

ganizations. Such a list can then be divided, even on a gross basis, into categories similar to those in Fig. 8.

The desired services that are deemed incompatible should then be evaluated more carefully. In most cases, only if the need for such services is given a high priority should the system design be structured to include them. Otherwise, the user should recognize the problems involved and look elsewhere for the service. The telephone network, for example, could well offer a better alternative. Another approach might be a small, "dedicated" network designed for these special services, independent of the main cable system.

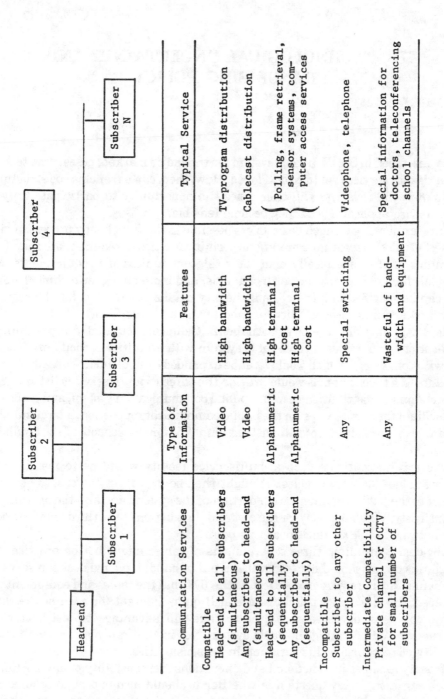

Communication Services	Type of Information	Features	Typical Service
Compatible			
Head-end to all subscribers (simultaneous)	Video	High bandwidth	TV-program distribution
Any subscriber to head-end (simultaneous)	Video	High bandwidth	Cablecast distribution
Head-end to all subscribers (sequentially)	Alphanumeric	High terminal cost	Polling, frame retrieval, sensor systems, computer access services
Any subscriber to head-end (sequentially)	Alphanumeric	High terminal cost	
Incompatible			
Subscriber to any other subscriber	Any	Special switching	Videophone, telephone
Intermediate Compatibility Private channel or CCTV for small number of subscribers	Any	Wasteful of band-width and equipment	Special information for doctors, teleconferencing school channels

Fig. 8—Cable-compatible services

III. TECHNOLOGICAL UNCERTAINTY AND FRANCHISING POLICY

As discussed in Sec. II, a cable system for the major markets poses some technological risk for services yet to be developed. How, then, can a franchise be structured to minimize the risk, and still offer cities the opportunity to participate fully in public-service applications as they become feasible?

Furthermore, how much flexibility is possible in the franchising process in view of the FCC rules? Recent interpretations permit the cities to require more than FCC minimums, if they specifically plan their telecommunications requirements. For example, channel capacity above 20 channels, and more stringent technical standards, can be specified in a franchise if the city details the reasons for the requirements and assumes responsibility for monitoring and compliance.

As a case in point, the city of Anaheim, California, circulated a preliminary specification for a proposed three-cable system with an initial capacity of 45 channels (with converters), which would later be expanded to an ultimate capacity of over 70 channels. These channels would provide the three kinds of services listed in Sec. I: entertainment distribution network, public communication system, and transmission medium for new services. In the public communication portion, a large number of channels were to be devoted to instructional TV for the schools of the Anaheim district.

One way to construct such a multiservice facility would be to install three separate cables (or sets of cables). It might then be logical for the city to bear the full cost of the public communication portion of the system, and also the operational responsibility. Who would pay for the cost and jurisdiction of the third "new service" cable network would be unclear at this point.

Obviously, installing three separate sets of cables might not be practical (although this option should not be dismissed out of hand) if a single set can serve all present and future requirements. Despite the fact that the cables and equipment can be shared, the city's responsibilities are still quite different for each of the three "subsystems." An objective should be to take full advantage of cost savings in constructing and operating the cable system, yet in the franchise provide for the flexibility of recognizing this variation in responsibilities.

If a city assumes jurisdictional and operational responsibility of any portion of the cable system, a key question is whether it should also expect to assume the financial responsibility. There is a distinct limit to the extent of the "gratuitous" services or capability that the franchise holder can be expected to absorb. A franchising authority may request certain public-service concessions in exchange for granting the franchise, but it is not realistic to assume that the cable system can, without some form of public subsidy, make up for all past, present, or future inadequacies of existing telecommunication services.

This question is particularly critical for any extensive use of the cable system by educational, health care, public safety and social services organizations, or what generally can be called "public institutional services." These are not subscriber-oriented, and if they are subsidized by subscriber revenues, they would seem to present an economic burden to both the subscribers and the cable system operators. In these applications, public financing would seem to be no more than justified.

An ideal franchising approach would involve all of the potential users of the cable system drawing up a comprehensive long-range plan, detailing their objectives and the required system capability to meet those objectives. These plans could then be combined with the available cable technology and optimized on a cost-effectiveness basis, and an answer could be given to the question of what kind of system is needed.

Unfortunately, this idealized approach is usually not practical. Most public agencies have insufficient familiarity with broadband telecommunication networks, typified by cable systems, to formulate such plans at this time. Nongovernmental community groups have less background in this area. In either case, some years may elapse before even the objective of such a plan can be determined.

Secondly, it is not yet clear who will pay for nonbroadcast services. Cities may eventually accept certain public communications as municipal responsibilities, in the same manner as providing police and fire department services, but it will be some time before they are established as budget items. Other new services may require some form of additional subscriber payment.

For these reasons, a more practicable approach is to concentrate on the technology, since this is more definable, and specify this technology in such a way that it permits a cable system installation to proceed, yet leaves the greatest flexibility in the areas of present uncertainty. If, additionally, the system is designed in a modular way, then it can accommodate more new services as they evolve, at a lower incremental cost.

It must be realized that modularity and flexibility are not achieved without penalty. Two of the most obvious penalties are a requirement for the franchising agency to plan the conceptual design of the system, and an increase in initial system cost to achieve the desired modularity.

Each community must determine whether these increased initial costs (usually passed on to the system's users) are justified either by achieving a wider range of services, or reducing the cost of future system expansion and modification. This determination cannot be easily made on a generalized basis applicable to all communities; it must take into account the specific characteristics and needs of the community in question.

DESIGN CONSTRAINTS

To achieve the stated objectives, the conceptual design tradeoffs must be evaluated carefully to determine what penalties would be paid if the cable system were constrained to reduce technical risk and increase flexibility.

It is emphasized that each major-market community must be evaluated on its own merits, since there generally are enough differences in local priorities (not to mention important geographic differences) to preclude applying any design approach in a "cookbook" manner.[1] The discussion that follows, therefore, is intended more as a generalized guide than as a definitive solution. Specifically, each of the areas of technical uncertainty in Sec. II is examined with the idea of arriving at a design approach that can be incorporated into the franchise requirements.

Cable technology is changing very rapidly, and a technical decision made in 1973 may not always appear to be the best one in, say, 1975 or 1980. Nevertheless, this possibility cannot be avoided, if the purpose is to begin the franchising process in 1973.

From today's viewpoint, the following six items (in the sequence described in Sec. II) might be typical design constraints for cable system. They are illustrative only, and are not intended to pertain equally to all cities.

1. A table of technical standards, test and compliance procedures, and corrective and penalty provisions would be compiled in detailed form for initial services, and in procedural or milestone form for future services.

2. Since set-top converters are still not as trouble-free as could be desired, two (or more) cables might be required for downstream transmission in order to initially supply 20 channels without converters. (This depends greatly, as noted previously, on the number of strong broadcast stations in the area. If there are from 4 to 7, a strong economic argument may favor the converter.)

Obviously, there is room for argument both on this judgment and also on the optimum number of cables. If two provide spare capacity, three would obviously provide more, at higher cost.[2] The decision involves some knowledge of the forecast trends in technology, plus a preliminary analysis of community priorities.

Again, the tradeoffs must be analyzed with care. It is fairly simple to calculate how much hardware and installation cost is added by requiring two downstream cables rather than one. It is less clear (and consequently requires more technological background) how much might be saved, for example, in service-call costs by delaying the early use of converters in the system.

3. Since two-way communications on the same cable carry a higher risk, and also limit the upstream capacity, the system design specifications might require a separate cable for reverse communications only.

To arrive at this decision, the higher initial cost (perhaps 10 to 20 percent higher than a single-cable installation for the cable alone, and ranging up to 30 to 40 percent higher if amplifiers and auxiliary equipment are included) must be weighed

---

[1] This is even true within the same major-market community. Beverly Hills, for example, lies within the Los Angeles market area, but as a separate city, it draws up its franchise independently. Its priorities may be very different from those of Watts.

[2] Indeed, a three-cable system in which one cable provides the "FCC required package" with the use of converters, and the other two cables form an inbound and outbound highway for two-way institutional, point-to-point, and interconnection services, appears to be an excellent approach in many cases.

against the anticipated importance of two-way services, and the estimates of required bandwidth. The impact of higher initial cost on system operations, subscriber rates, and revenues should be considered explicitly to determine if the economic viability of the system is endangered.

4. The system design might stress both reliability and maintainability as part of the initial concept and not as afterthoughts to be applied retroactively. Thus, the following techniques could possibly be considered for inclusion as franchise requirements:

- Provision of test points throughout the system.
- The use of redundancy of the most failure-prone components (usually the "active" elements such as amplifiers).
- Backup or emergency power supplies.
- Protection from tampering (for high-security applications such as police alarm systems).
- Qualification and inspection testing of components, aging, and "burn-in" of electronic equipment prior to installation.
- Imposition of mean-time-between-failure and maximum-time-to-repair criteria.
- Ease of access for repair or replacement.
- Documentation of test and repair procedures.
- Training of maintenance personnel.

It is not be inferred that all of the above are necessarily worth the cost and effort, but they should be evaluated before a franchise is granted, since they would have value only if included in the original system design

5. System-interconnection requirements might be specified from a functional point of view, but in sufficient detail to permit estimating cost tradeoffs. This would include number of channels inbound and outbound for each cable system, requirements for subdistricting within any single franchise area, assignment of standardized frequencies for interconnection, and electrical interface requirements where the connections are made.

6. Special equipment for new services should not be required before the market for those services has been defined, but the system should be initially designed to accommodate the added equipment both physically and electrically, so far as possible.

This constraint includes fairly clear examples (e.g., do not install upstream amplifiers until actually needed) and some not so clear (e.g., how to accommodate computer-controlled digital equipment in the cable system without a firm set of standards or component designs). Consequently, both familiarity with developmental and pilot experiments in cable technology and knowledge of how to extrapolate their future impact on existing cable systems are required. Some of this familiarity can be gained from the available cable system literature. Beyond that, franchising authorities will need outside assistance during the pre-franchising planning stages.

Apart from such generalized design constraints, each municipality or region will have certain specific constraints that must also be considered (e.g., the eventual incorporation of existing telecommunication services and facilities into the new cable network). The general and specific constraints together form the base from which the modularized system design can be developed.

AN EXAMPLE OF MODULARIZED SYSTEM DESIGN

When the generalized objective of achieving a cable system design flexible and modular enough to permit growth with reasonable cost is applied to the case of a specific community, the first task is to relate community needs and requirements (so far as they are known) with cable technology and costs. From this relation, which often is an iterative process, a set of design constraints are formed that resemble those of the previous section but are "tailored" for that particular community. These in turn lead to a system configuration that best fits those constraints.

Very often, more than one configuration can be reasonably considered desirable and the choice is difficult, depending primarily on which features are given priority.

As an example, it has been noted that a popular design at present is to install a two-cable system, with one trunk cable "active" and the second trunk a "shadow." The active cable is fully connected (e.g., amplifiers, taps, drops, etc.) for downstream transmission only. With converters, some 25 to 35 channels can be provided, meeting current FCC requirements.

The shadow trunk cable usually has no auxiliary components connected but has provision (e.g., empty housing) for later installation of amplifiers and other elements. If more downstream channels are required, the shadow cable can supply these. Similarly, when upstream transmission becomes needed for two-way services, the shadow cable can be used in a simultaneous two-way mode by either the "low-band" or "mid-split" technique described in Sec. II.

Obviously, this approach is both flexible and modular, and has the virtue of low initial cost. It does impose some technical risk, however, both explicit and implicit. First, the problem of interference between upstream and downstream communications on the same cable is postponed rather than avoided (which has some benefit, since it is to be expected that techniques to minimize interference effects will improve, and the problem may be substantially solved by the time the shadow cable is actually needed for new services).

Second, the question of reliability for two-way services (which, as noted, is much more critical) may not receive appropriate consideration. The reason for having the shadow trunk is more to provide a relatively low-cost vehicle for future expansion, than to design one highest-reliability transmission medium for two-way services.

An alternative to the shadow trunk is to provide a completely separate trunk cable for reverse transmission, which means trading higher initial cost for lower technological risk. For illustration only, this latter approach is chosen to suggest how system expansion can occur in a flexible, modular fashion, and also how this flexibility can be reflected in the structure of the franchise itself.

Figures 9 through 12 show, in simplified form, the incremental expansion of the illustrative cable system.

Figure 9 shows a "basic" cable TV system capable of one-way, TV-type communications. Using the previously mentioned generalized design constraints as a guide, a dual-trunk, dual-feeder cable system has been selected, without converters initially. Each of the two downstream cables can deliver a maximum of 12 channels to the TV receiver, and an "A-B" switch for selecting either cable A or B is mounted on top of, or adjacent to, the receiver. Thus, a nominal 24 channels is available, but, as noted, this must be reduced by the number of locally broadcast signals.

If there are two local broadcast stations, only 10 channel positions can be used

199

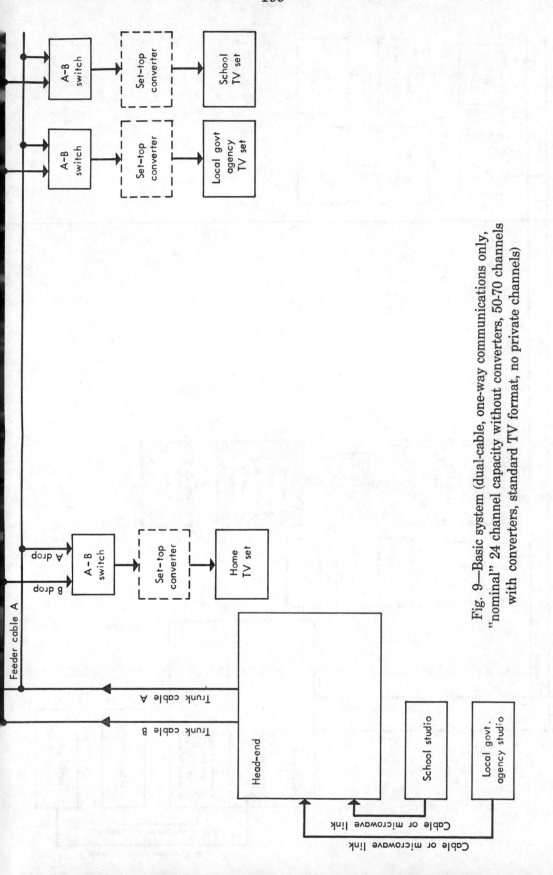

Fig. 9—Basic system (dual-cable, one-way communications only, "nominal" 24 channel capacity without converters, 50-70 channels with converters, standard TV format, no private channels)

200

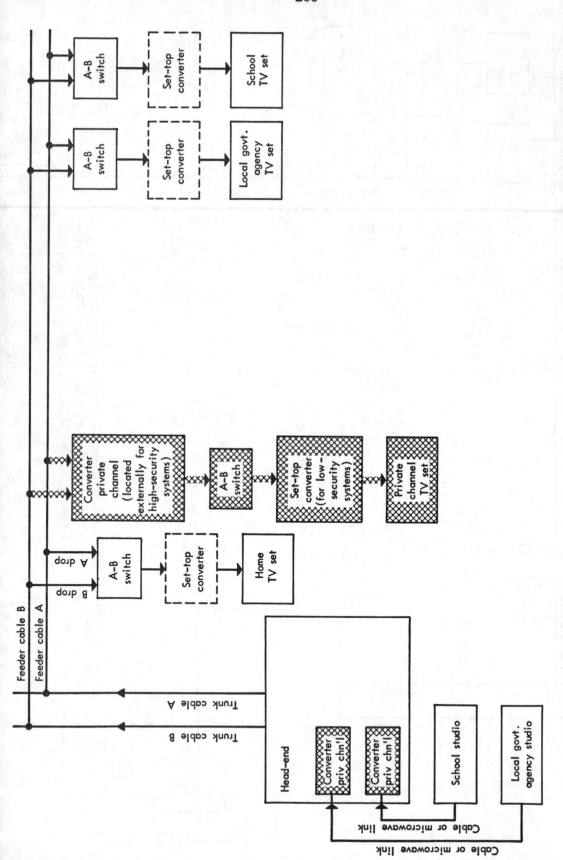

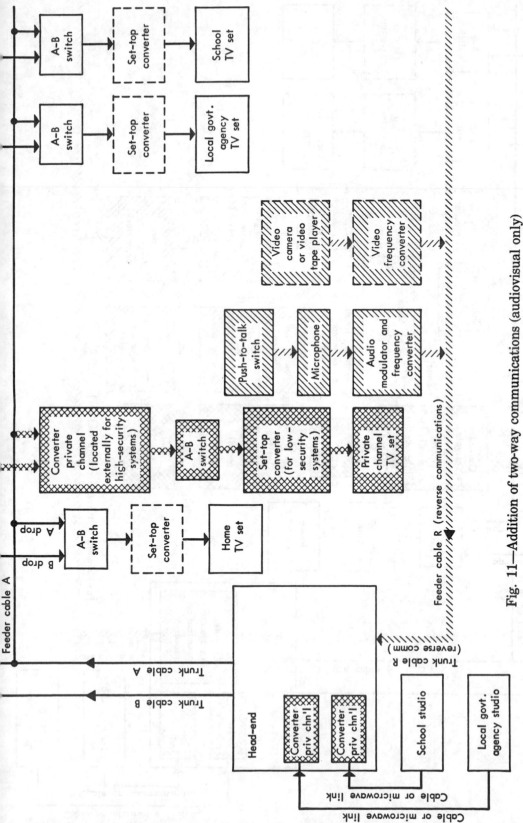

Fig. 11—Addition of two-way communications (audiovisual only)

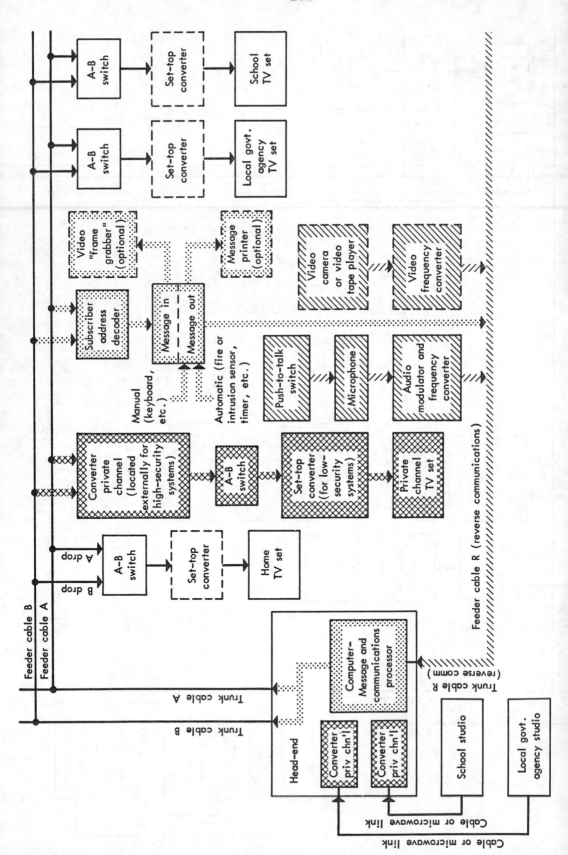

at the TV receiver, and the system has a 20-channel capacity for Class I (TV-type) signals. (If 7 strong local broadcast signals exist, only 5 channel positions are available per cable, necessitating more cables to provide 20 Class I channels. This may make a multicable approach unwieldy.)

Figure 9 shows the basic system with the addition of a converter (broken lines) at each TV receiver. This incremental expansion can be accomplished at any time and is dictated by two considerations: (1) when more channels are needed; and (2) when converter performance and reliability are considered adequate.

It may not be necessary to add converters at all TV receiver locations. If, for example, a limited-audience service is contemplated (e.g., continuing medical education for doctors and nurses), only those subscribing to, or authorized for, this service would require the converter. Thus the cost would be borne by the recipients of the benefits.

Figure 10 illustrates the extension of this concept, showing the incremental added equipment permitting "private-channel" cablecasting. Such private channels could either be commercial (i.e., pay-TV) or governmental (municipal agency teleconferencing). In each case, the private channel signal is coded or converted before distribution over the cable so that only locations with an appropriate decoding component can receive the information.

It should also be recognized that at least two levels of privacy may be required. For a pay-TV entertainment channel, it may be acceptable to have the decoding device physically located at the TV receiver, but for transmitting medical or welfare records or other sensitive data, access to the decoding equipment would have to be restricted to prevent unauthorized tampering. In this case, such equipment might be installed outside the subscriber's area, either on the poles or at a central facility.

The essential point is that the system should be able to accept either or both private-channel techniques without undue cost in the future. Thus, the planning of the initial installation should include provision for this later incremental expansion.

Figure 11 shows the addition of two-way communications, which is limited to only audiovisual types of messages. Here the separate cable R, which is postulated as part of the initial installation, is activated for reverse transmission. (The amplifiers for cable R, for example, need not be purchased and installed until the need for such reverse communications has been defined. Thus the only initial cost involved is that of including cable R, which is relatively moderate and definable.)

For upstream audio messages, each location that originates a message will require the addition of equipment shown in the solid lines in Fig. 11. This is similar to equipment used in any voice-only teleconferencing activity. Technically, the only problem is the interface to the cable, which must be designed so that a large number of tap-in points will not cause excessive noise or interference. This is not expected to be a major barrier, but to date no large cable system has been built with many upstream signal-origination locations.

The capability for upstream video signals is provided by adding the broken-line equipment shown in Fig. 11. Again, in principle, there is no limit to the number or locations of origination points (subject to cable bandwidth), but multiple connection to the cable without causing interference still remains a design problem.

Finally, Fig. 12 illustrates the added equipment necessary to permit two-way alphanumeric communications, usually computer-controlled. This capability will permit most of the "blue sky" cable applications that have received publicity over the past 15 years (audience polling, electronic mail, automated shopping information retrieval, sensor-activated alarm systems, and the like).

This capability is now being developed and demonstrated,[3] but many difficulties must be overcome before such feasibility demonstration becomes an economically viable possibility. Currently, the terminal equipment required at the home is expensive, necessitating perhaps an order-of-magnitude cost reduction. The software required for any new service is also costly, and time-consuming to prepare. Furthermore, no determination has been made of either the cost of specific new services or who would be expected to bear the cost. Perhaps 2 to 5 years will be required to evaluate pilot systems, services, and costs before planning to implement the capability of Fig. 12 can commence for any major market.

It would be difficult to determine which will occur first for most cable systems: two-way video services or data-response services. Shared upstream video, while less complex technically, requires more planning (and budgeting) by city agencies or educational districts, since most applications would be expected in the public-service area. On the other hand, data-response services are being implemented on a pilot basis, and if the house terminal cost can be reduced sufficiently, they offer the prospect of many exciting new applications. Since it is problematical which capability will be included first, the franchise structure should be capable of dealing with either, regardless of the sequence.

Accompanying Figs. 9 through 12 is the chart of Table 2, which summarizes the capability and range of services made possible by each growth increment in the cable system. This is useful, if prepared in more detail, as both a checklist and stimulant for municipal agencies and community organizations in determining the terms of the franchise. Such preliminary review, even if not conclusive, is valuable in determining the relative importance of various new services as well as the degree of definition included in the franchise.

As a typical example, officials of a local school district can use the cable system for the following:

- Cablecasting instructional TV programs over an assigned channel, without regard for privacy (Fig. 9).
- Cablecasting instructional TV, with viewing restricted to designated schools or classes (Fig. 10).
- Cablecasting instructional TV, with provision for student talk-back (Fig. 11).

If the cable system were planned as shown in Figs. 9 through 12 with the detail necessary for the specific municipality (e.g., number and location of schools, classrooms, etc.), and included estimated costs and implementation difficulties for each expansion increment, the school district would then have a reasonable basis for making tradeoff evaluations, and could determine whether a desired educational capability is worth the expenditure.

It is to be hoped that many public agencies (if not all) and interested community organizations will perform a similar analysis, or that a single group will be delegated

---

[3] The Hughes-Theta Subscriber Response System scheduled for demonstration on the El Segundo, California, cable system is an example. Some half-dozen other installations by various manufacturers are also either under construction or in the design stage at present.

Table 2

SUMMARY OF CAPABILITIES AND RANGE OF SERVICES ILLUSTRATED IN FIGS. 9 THROUGH 12

Cable-System Configuration	Channel Capacity	Functional Capability	Typical Service
	Figure 9		
Basic cable TV system (dual-cable construction)	(a) 24 channels without set-top converters (12 channels on each cable; however, this theoretical maximum must be reduced by the number of strong broadcast TV channels, since off-the-air interference will prevent those channels from being used for cablecasting) (b) 50 to 70 channels with converters	(a) One-way communications only (b) Standard TV format required (c) No private channels (i.e., all subscribers can receive any cablecast program) (d) Programs that originate at various locations (schools, city government offices, etc.) can be cablecast if connected to the system distribution center(s)	(a) Distribution of commercial TV programs from the headend to subscribers (b) Distribution of locally originated programs (e.g., cablecast by CATV operator or public access channel programming) (c) Distribution of remotely originated programs through links to cable-system distribution center(s) (e.g., from school or local government studios)
	Figure 10		
Basic system with private-channel capability	Same as (a) and (b) above	(e) In addition to (a) through (d) above, any channel can be made "private" (i.e., its programs restricted only to authorized reception points that are furnished with appropriate special-conversion equipment)	(d) In addition to (a) through (c) above, private channel cablecasting permits such services as (1) School programming restricted only to certain (or all) schools (2) Local government programs restricted to designated government agencies (3) Specialized programming to subscribers (e.g., medical information to doctors, etc., pay-TV channels)
	Figure 11		
Basic system with private-channel capability, two-way audio-visual communications (third cable added for reverse, or upstream communications)	(a) For downstream communications, same as (a) and (b) (b) For upstream communications, equivalent bandwidth of up to 35 TV channels. Since many upstream signals are not destined for standard TV receivers, the limitations of 12 switchable channels and assigned frequencies do not apply	(f) In addition to (a)–(e) above, the solid lines of Fig. 11 permit interactive TV with audio-only talk-back (g) The dashed lines of Fig. 11 permit upstream video transmission as well	(e) Solid lines of Fig. 11 permit (1) Educational TV with student response and inquiry (2) Any form of teleconferencing where the video (picture information) is generated at only one location (f) Dashed lines of Fig. 11 permit teleconferencing with video originating from multiple locations (on channel required for each
	Figure 12		
Basic system with private-channel capability, two-way audiovisual communications, and two-way computer-controlled alphanumeric communications	(a) Same as (a) and (b) for Fig. 11	(h) The addition of Fig. 12 equipment permits each subscriber to be polled individually under computer control, and provides any specific services the subscriber requests	(g) Audience polling, sensor monitoring networks, automated transaction processing (shopping, banking, etc.), information retrieval, and computer time-sharing

to perform this task for all. No matter how crude or preliminary the inputs from such an analysis, they would be more specific and useful to the franchising authority than what is generally available now.

The tasks of analyzing public-agency objectives vis-a-vis cable-system technology and costs are substantial. In many cases, they will require additions to municipal and county staff, perhaps initially on an ad hoc basis, but thereafter on a permanent, ongoing basis.

RELATION BETWEEN SYSTEM DESIGN AND FRANCHISE STRUCTURE

There are two major reasons for guiding the design of the cable system along modularized concepts. The first is to permit technological expansion of the system without catastrophic retrofit or obsolescence penalties. The second reason, perhaps more important, is to be able also to structure the *franchise provisions* so that such expansion occurs under reasonably clear and controlled conditions. This is certainly no easy task, but it appears necessary to make the attempt if (1) the franchising authority wishes to grant a cable TV franchise without an indefinite delay; and (2) it also wishes not to abdicate, or seriously dilute, responsibility and control in the new services area.

The latter point deserves some amplification. Almost all franchises, particularly recent ones, include some terminology that gives the impression of maintaining responsibility and control. The franchise may require the cable TV system operator to "keep up with the state of the art," or to "have the capability to add two-way communications when required," or to "provide more government or educational channels at a later date."

There is some validity to deliberately stating such requirements in broad, qualitative language. It becomes much easier to allow the franchise to begin sooner, and it frees the franchising city or county from some of the monitoring, validation, and compliance procedures that would be necessary if specific design specifications were spelled out. The franchising authority has wide flexibility to request system expansion or upgrading, and to negotiate the terms and conditions. The power of franchise cancellation acts as an implicit "big stick" during the negotiations.

The penalties involved with using such broad language are serious, however. First, the initial system design concept and the methods to be used in expanding the system in the future are left completely to the discretion of the franchisee. Even with the best of intentions, the operator of the cable system cannot be expected to use the same tradeoff and evaluation criteria as the officials of a city or county government might use. Consequently, the designs almost always will be oriented toward minimizing initial installation costs, which, in turn, will usually cause much more expense and difficulty in achieving orderly growth. The franchising authority may become aware of this only when an incremental expansion is requested, perhaps some years after the franchise is granted.

As one example, the FCC now requires bidirectional capability; and single-cable systems, as was shown, can comply with the letter of this requirement. If the city were to need 5 or 6 video upstream channels several years later, however, it might find the system incapable of responding to this requirement without extensive re-trofitting. It may well be that retrofitting later will not cost more than paying for the capability initially, but this tradeoff should be a prefranchise consideration.

With the broad language approach, the success of the city or county in achieving its growth objectives depends heavily on such subjective (and variable) factors as the personal capability of its negotiators. The questions of how much the franchising authority is legally entitled to, how much it will cost, who will bear the cost, etc., become subject to a wide range of ad hoc interpretation.

If these problems are to be minimized, the franchise provisions that govern expansion of the system must be related to the initial design concept, with each required growth increment based on the anticipated technology and costs.

A rudimentary illustration of how to do this is provided in Table 3. Each configuration shown in Figs. 9 through 12 is evaluated in terms of the current technological status of the equipment required, cost estimates, and a forecast of when the particular system capability might either be needed or in demand. For any specific geographic area, other factors could also be included. The structure of the franchise can be developed from this chart and similar support data. As shown, the Fig. 9 configuration can be completely defined, with respect to the terms and conditions and also the responsibilities. Proof-of-compliance procedures and appropriate penalties for noncompliance also can be defined.

For the future expansion increments of the system, the franchise need not be (and indeed, realistically cannot be) that comprehensive. Instead, franchise milestones can be established that determine when each increment shall be considered for addition, and, when, in fact, it is to be added. Furthermore, each milestone has the option of carrying or not carrying with it its own new negotiation, not only with respect to who pays the cost of the system expansion, but also who carries the operating responsibility for a new service.

As a hypothetical case, assume that the city fire department progresses to the point of wanting to replace street alarm boxes with a citywide cable fire-alarm system. Also, assume that at the time the plans and objectives become firm, the technology makes the new service possible (e.g., low-cost reliable heat sensors, system security, techniques that minimize false alarms, etc.).

Many questions will arise, including:

- Who pays for expanding the cable system to achieve a 100-percent citywide connection?
- Who assigns and controls the cable bandwidth necessary?
- Who physically operates and maintains the cable channel and ancillary equipment for this service?
- Who pays for undergrounding, if such a service demands it?

Even these few illustrative questions indicate that implementation will be a highly complex task. The service is public, and presumably the city will wish to control it as it would any other municipal service; yet it is intertwined with a commercial venture.

No advance planning can eliminate all the obstacles to be encountered. But at least a good start can be made toward negotiating the expansion of the cable system to include this desirable service, if the franchise is initially structured to permit the city to:

Table 3

EVALUATION OF THE CONFIGURATIONS IN FIGS. 9 THROUGH 12

Figure	Status of Technology	Cost Range	Cost Responsibility	Need for Capability	Possible Franchise Action
9 (without converters)	Available now	Within traditional monthly subscriber charges	Subscriber	Required as part of initial installation	Can be completely defined
9 (with converters)	Available now with reservations as to converter reliability and performance	Converter cost of about $20-40; adds $1-2 per month to basic charge	Subscriber	May or may not be needed initially, depending on local broadcast signals. Increased numbers of channels probably not needed for 3-5 years	Franchise milestone can be established, including terms and conditions to add this capability
10 (private channel)	Developing rapidly, but not yet demonstrated in large systems	Cost of home decoders about $50-100	Subscriber will pay for commercial service; uncertain for public service	Pay-TV may occur in 1-3 years; other applications in 3-8 years	Two franchise milestones may be appropriate: (1) for commerical applications, and (2) for public applications
11 (upstream audiovideo)	Developing rapidly, but not yet demonstrated in large systems	For audio, $50-200 per origination point; for video, $500-3000 for black & white, $1000-10,000 for color	Uncertain	Requires 1-3 years of public agency planning; capability may be needed in 3-7 years	Two franchise milestones may be appropriate: (1) for commercial applications, and (2) for public applications
12 (two-way data communications)	Experimental, pilot demonstrations on any sizable scale scheduled for 1972-1973	For home terminal, $100-1000; central computer costs for large-scale information retrieval may double this	Uncertain	Requires both planning and extensive funding of equipment and programs; capability may be needed in 3-10 years	May be too nebulous to include in initial franchise. It would probably be more appropriate to license proposed new service on an individual basis

- Compel the fire-alarm implementation, granting a fair price to the cable operator for any new costs assumed (the "fair price" might be based on prior costs and profitability, or some other defined formula);
- Negotiate a formula for city use of sufficient bandwidth to make such a service possible (perhaps by leasing a block of channels under specified conditions); and
- Treat this portion of the system as a municipal service against which public tax revenues can be applied.

In contrast, if such a service were desired under the terms of present franchises, it would be almost impossible even to guess at the outcome. The probabilities are that in most cases, such new services simply would not be implemented over the cable network due to the complexities involved.

Although it does not solve every problem, modularization of both the system design and the franchise, on an interdependent basis, offers an orderly path to system growth and the introduction of new services.

IV. APPROACH TO FRANCHISE DECISIONS

It is evident even from the limited discussions of Secs. II and III that a major-market cable system is a complex, expensive communication network. The cost for a cable television system for the city of Houston, the 15th-ranked market in the United States, was recently estimated at about $50 million without two-way capability [5]. (Houston has a population of 1.8 million and covers a 500-square-mile area.)

If this were simply a commercial venture, the risk and the need for careful planning would be the responsibility only of the franchise holders. Since the cable system already involves substantial local government participation, however, and is expected to require more in the future, the system definition and planning must be shared if the public-service aspect is to be awarded proper emphasis.

Planning a multimillion-dollar network is no mean task, particularly when much of the system design is for future services not yet defined. The interdisciplinary expertise required may exceed the present staff capability of even the largest cities.

The penalty for not sharing the preliminary planning, however, can be drastic. The ability to add services at a later date can be foreclosed completely by the nature of the initial design, or made so costly and difficult that the services are not worth implementing. In view of the extensive publicity given to cable systems as a cornucopia of beneficial, community-oriented new services, it would be unfortunate if the major-market systems did not fulfill their technological potential.

To avoid this, and to be able to use the modular approach proposed in Sec. III, the franchising agency must accept a considerable prefranchise planning responsibility, requiring both an outlay of "front-end" funds and competent planning personnel.

THE PLANNING PROCESS

Figure 13 diagrams a chronological flow of the steps that might be taken as part of a major-market-system planning and franchising process. Phases 1 through 3 are the planning tasks that would occur prior to drawing up a franchise.

Phase 1 is a familiarization period during which representatives of local government agencies and interested community organizations acquire information on the nature and capability of cable TV systems. People's familiarity with cable systems

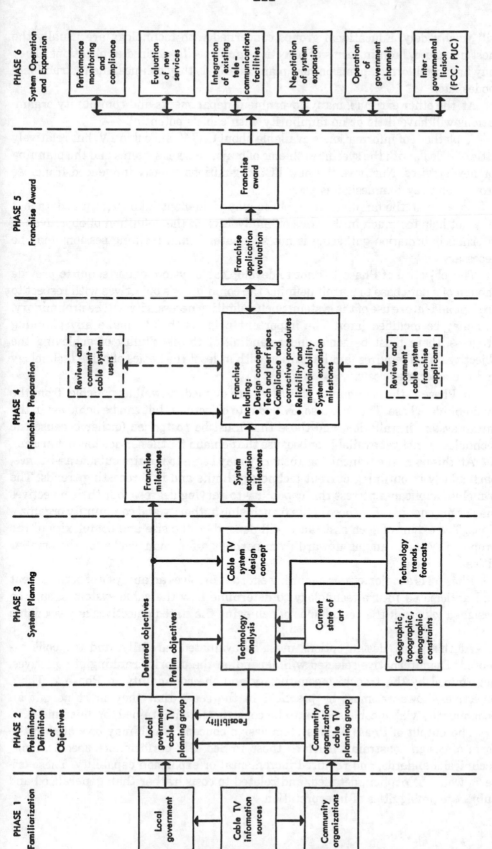

Fig. 13—Cable TV system franchising

will vary widely. Educational groups concerned with technology are usually the most advanced, both in terms of objectives and specific desired applications. They may have well-formed plans on how many channels they can use, the instructional modes, etc.

At the other extreme, many government departments and community organizations will have little or no familiarity with cable's potential.

A plethora of information is available about CATV and cable TV, but relatively little of it deals with the specific problems of major-market systems and the planning for new services. Nor have the new FCC regulations greatly influenced franchise provisions or system design as yet.

As a result, the familiarization phase may yield plentiful background data, but may not help too much in the tradeoff analysis or in the definition of objectives. If informal information gathering is not adequate, formal training sessions may be necessary.

The objective of Phase 1 is not to develop cable system experts, but to provide enough of a data base to permit defining the community's objectives with respect to present and future use of the system. Such definition necessarily will be preliminary, and may be modified later. The important point is that to make any planning progress, there must be some understanding of the feasibility of achieving the objectives. This requires familiarization with at least the basics of cable technology and the associated costs.

In Phase 2, the community objectives are defined, as well as possible, from the available data base. They are also divided into objectives that can be achieved in the initial system installation, and those that must be postponed (either because the technology is not yet available or because the demand for the service has not arisen).

At this point the franchising authority may begin to require outside assistance, particularly in outlining current technology limits and in forecasting trends. The franchise applicants possess this capability to varying degrees, but their objectives may or may not parallel those of the franchising authority and community organizations. The extent of such assistance will depend on the size and complexity of the proposed system, ranging upward from perhaps a few man-weeks for the smaller cities.

Phase 3 is a major system-planning effort. It involves an analysis of both current and anticipated future technology to determine how the cable system should be designed to permit the best chance of achieving the stated objectives at reasonable cost.

At this point, design options must be evaluated carefully, and tradeoffs explored. The alternative selected will depend on the priority ranking of objectives, determined by the franchising authority and the community in Phase 2. These objectives, however, must be practical in the sense that they must permit an economically viable cable system to be constructed, maintained, and expanded.

The output of Phase 3 is a system-design concept, which may be stated in the form of design constraints (such as those in Sec. III), performance specifications, technical standards, and required incremental or expansion capability. These criteria would be explicit, detailed, and related to costs, rather than generalized and subject to ambiguities of interpretation.

If successful, this design output will achieve two goals:

1. Provide the franchising authority with the basis for a cable-system design that is both feasible and flexible to expansion along paths the community deems most desirable.
2. Standardize, to a large extent, the applications submitted for franchise consideration. This will minimize the "apples against pears" comparison that has long been a feature of cable-system franchise applications, and simplify the task of the franchising authority in selecting the most favorable applicant.

THE FRANCHISING PROCESS

Phase 4, the preparation of the franchise documents, incorporates the results of Phase 3. The design concept becomes part of the franchise contract. Compliance testing and corrective and penalty provisions are included, based upon possible failure modes of the specific system. Reliability and maintainability requirements are also specified.

Anticipated expansion increments of the physical system are tied to franchise provisions that either permit new negotiations or follow a prescribed set of expansion ground rules. These may specify procedures to be followed (e.g., the system operator must provide the new service, the city must compensate the operator accordingly, etc.) to assure that a clear path exists for achieving the new services. They may defer the questions of licensing, operation, and control of new services, to be agreed upon when the expansion is desired.

Drafts of the proposed franchise should be reviewed by prospective system users and also by a representative sample of cable-system operators (both applicants and, more impartially, some who are not applying for that particular franchise). The critique resulting from the review may lead to modification of the franchise.

Once the franchise is in final form, the city or county is ready to accept and evaluate applications (Phase 5). As noted, one of the most beneficial by-products of this procedure is that all applicants will now be bidding to essentially the same design (although applicants should remain free to exercise creativity in offering alternative options), and it becomes much simpler to evaluate the various bids against each other. The award can be made on the basis of preannounced, weighted criteria.

Considering cost alone, some of the benefits of prefranchise planning are realized in reducing the time and expenditure of the applicant evaluation process, which is typically a crucial and thorny task.

The staff that a franchising authority may require for Phases 4 and 5 will include the cognizant government officials of the city or county (e.g., City Manager or Mayor's office, staff of the City Council or Board of Supervisors, City Attorney, etc.). They should be assisted by technological expertise, so that the franchise provisions are legally and appropriately related to the design requirements of Phase 3.

MAINTAINING A CONTINUING EFFORT

When the franchise has been awarded, the city's or county's task is usually just beginning. Phase 6 of Fig. 13 illustrates some of the functions that should be established and maintained on a continuing basis.

For a small community, these functions usually will be performed by the present governmental staff. An advisory group or committee (usually unpaid) might be appointed that is representative of various community interests (education, health care, communications media, social services), to redefine objectives for cable services over periods of time. The technical requirements of monitoring system performance might be achieved by witnessing tests made by the franchise holder's employees or by assigning this duty to the City Engineer or City Electrician.

For communities of perhaps more than 250,000 in population, permanent additions to the city or county staff may be advisable. If there is already a city Director of Communications, for example, he may have to add personnel with specific responsibility for cable-system monitoring and planning. If such an office does not exist, the cable system will probably justify its authorization.

At the major-market level, where the population might be over 1,000,000, the tasks may warrant the establishment of a high-level Office of Telecommunications, which would perhaps be directly responsible to elected officials, such as the Mayor, City Council, or Board of Supervisors.[1]

At the municipal or county level, this office would not only parallel the strategic and policy functions of the federal Office of Telecommunications Policy but would also have the detailed responsibilities of system design, coordination of disparate communications facilities, and the development of both a near-term and long-range telecommunications plan.

One possible organizational framework for this office is illustrated in Fig. 14. It includes four functional divisions:

1. Technical Performance and Standards,
2. System Planning and Development,
3. System Operations, and
4. Telecommunications Policy.

The first function is concerned with the immediate operational aspect of telecommunications systems already installed, or due to be installed (such as a franchised cable TV system).

System Planning and Development includes the responsibility for long-range telecommunications planning on an area-wide basis. First, the existing municipal or county systems and facilities (police, fire, civil defense, etc.) would be reviewed, and the desired capabilities determined for perhaps the next 5, 10, 15, and 20 years. Taking the anticipated availability of cable channels into account, a phased series of plans would be formulated, detailing the transition from present facilities to the planned future configurations. Various options and alternative systems would be analyzed on a cost-effectiveness basis.

The intent would not be to supersede the individual governmental departments' communications plans, but to provide the overall system coordination and technical

---

[1] At this writing, only New York City has established such an office.

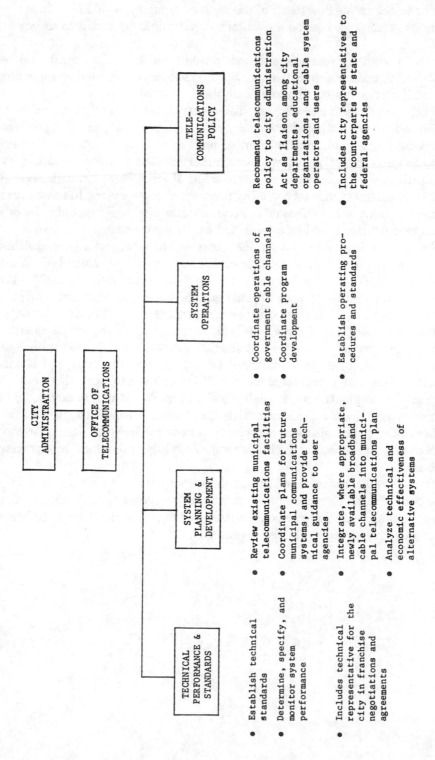

Fig. 14—Organizational framework for an Office of Telecommunications

expertise that would permit achieving those plans most effectively. One responsibility would be to determine how much of the desired capability could be satisfied by the system as structured, and, as a corollary, what might be added to satisfy all objectives.

The third function, System Operations, would coordinate the operational responsibility for all telecommunications services. The local-government channel now available on the cable system is a key example.

The function of the fourth division, Telecommunications Policy, would be to recommend policies permitting effective telecommunications planning and performance. This would include technically oriented policies such as allocation of channel space, system interconnection requirements, determination of two-way capability, etc. It would also be concerned with some nontechnical areas, however, such as usage of public-access channels and responsibility for costs of new services. In these areas, the recommendations of the Office of Telecommunications would be only one of a group of inputs; policies would be established at higher level.

Since many services and modes of operation will be new, initially established policies no doubt will be modified with time. Some areas will be particularly sensitive, both in terms of defining policy and carrying it out. An obvious example is the unique relationship that is still evolving between the cable-system operator and the city. There has been considerable policy experience between cities and regulated franchises (e.g., buses or taxis), but relatively little experience with joint commercial and municipal enterprises, where the cost-sharing structure remains as yet unclear. The details of the final relationship cannot be predicted at this time, but it does appear that a major and coordinated effort will be required to work them out.

The cost of assuming the responsibilities shown in Fig. 14 will undoubtedly be high. In many cases, it will exceed the anticipated franchise fees. If the justification for an Office of Telecommunications is spelled out in detail, however, it may provide a legitimate basis for requesting the FCC to approve a higher fee than the 3 percent now established.

In any event, the influence of the cable system upon the community may become so pervasive, and its value to the community so great, that the franchising authority cannot abdicate those responsibilities, and the costs may have to be borne whether or not they outweigh the fee revenues.

Appendix A

FCC DEFINITION OF CLASSES OF CABLE TELEVISION CHANNELS

1. *Class I cable television channel.* A signaling path provided by a cable television system to relay, to subscriber terminals, television broadcast programs that are received off-the-air or are obtained by microwave or by direct connection to a television broadcast station.
2. *Class II cable television channel.* A signaling path provided by a cable television system to deliver, to subscriber terminals, television signals that are intended for reception by a television broadcast receiver without the use of an auxiliary decoding device and which signals are not involved in a broadcast transmission path.
3. *Class III cable television channel.* A signaling path provided by a cable television system to deliver, to subscriber terminals, signals that are intended for reception by equipment other than a television broadcast receiver or by a television broadcast receiver only when used with auxiliary decoding equipment.
4. *Class IV cable television channel.* A signaling path provided by a cable television system to transmit signals of any type from a subscriber terminal to another point in the cable television system.

Appendix B

FCC TECHNICAL STANDARDS

Subpart K—Technical Standards

§ 76.601 Performance tests.

(a) The operator of each cable television system shall be responsible for insuring that each such system is designed, installed, and operated in a manner that fully complies with the provisions of this subpart. Each system operator shall be prepared to show, on request by an authorized representative of the Commission, that the system does, in fact, comply with the rules.

(b) The operator of each cable television system shall maintain at its local office a current listing of the cable television channels which that system delivers to its subscribers and the station or stations whose signals are delivered on each Class I cable television channel, and shall specify for each subscriber the minimum visual signal level it maintains on each Class I cable television channel under normal operating conditions.

(c) The operator of each cable television system shall conduct complete performance tests of that system at least once each calendar year (at intervals not to exceed 14 months) and shall maintain the resulting test data on file at the system's local office for at least five (5) years. It shall be made available for inspection by the Commission on request. The performance tests shall be directed at determining the extent to which the system complies with all the technical standards set forth in § 76.605. The tests shall be made on each Class I cable television channel specified pursuant to paragraph (b) of this section, and shall include measurements made at no less than three widely separated points in the system, at least one of which is representative of terminals most distant from the system input in terms of cable distance. The measurements may be taken at convenient monitoring points in the cable network: *Provided,* That data shall

be included to relate the measured performance to the system performance as would be viewed from a nearby subscriber terminal. A description of intruments and procedure and a statement of the qualifications of the person performing the tests shall be included.

(d) Successful completion of the performance tests required by paragraph (c) of this section does not relieve the system of the obligation to comply with all pertinent technical standards at all subscriber terminals. Additional tests, repeat tests, or tests involving specified subscriber terminals may be required by the Commission in order to secure compliance with the technical standards.

(e) All of the provisions of this section shall become effective March 31, 1972.

§ 76.605 Technical standards.

(a) The following requirements apply to the performance of a cable television system as measured at any subscriber terminal with a matched termination, and to each of the Class I cable television channels in the system:

(1) The frequency boundaries of cable television channels delivered to subscriber terminals shall conform to those set forth in § 73.603(a) of this chapter: *Provided, however,* That on special application including an adequate showing of public interest, other channel arrangements may be approved.

(2) The frequency of the visual carrier shall be maintained 1.25 MHz±25 kHz above the lower boundary of the cable television channel, except that, in those systems that supply subscribers with a converter in order to facilitate delivery of cable television channels, the frequency of the visual carrier at the output of each such converter shall be maintained 1.25 MHz±250 kHz above the lower frequency boundary of the cable television channel.

(3) The frequency of the aural carrier shall be 4.5 MHz±1 kHz above the

218

frequency of the visual carrier.

(4) The visual signal level, across a terminating impedance which correctly matches the internal impedance of the cable system as viewed from the subscriber terminals, shall be not less than the following appropriate value:

Internal impedance:
    75 ohms.
    300 ohms.
Visual signal level:
    1 millivolt.
    2 millivolts.

(At other impedance values, the minimum visual signal level shall be $\sqrt{0.0133\ Z}$ millivolts, where Z is the appropriate impedance value.)

(5) The visual signal level on each channel shall not vary more than 12 decibels overall, and shall be maintained within

(i) 3 decibels of the visual signal level of any visual carrier within 6 MHz nominal frequency separation, and

(ii) 12 decibels of the visual signal level on any other channel, and

(iii) A maximum level such that signal degradation due to overload in the subscriber's receiver does not occur.

(6) The rms voltage of the aural signal shall be maintained between 13 and 17 decibels below the associated visual signal level.

(7) The peak-to-peak variation in visual signal level caused undesired low frequency disturbances (hum or repetitive transients) generated within the system, or by inadequate low frequency response, shall not exceed 5 percent of the visual signal level.

(8) The channel frequency response shall be within a range of ±2 decibels for all frequencies within −1 MHz and +4 MHz of the visual carrier frequency.

(9) The ratio of visual signal level to system noise, and of visual signal level to any undesired cochannel television signal operating on proper offset assignment, shall be not less than 36 decibels. This requirement is applicable to:

(i) Each signal which is delivered by a cable television system to subscribers within the predicted Grade B contour for that signal, or

(ii) Each signal which is first picked up within its predicted Grade B contour.

(10) The ratio of visual signal level to the rms amplitude of any coherent disturbances such as intermodulation products or discrete-frequency interfering signals not operating on proper offset assignments shall not be less than 46 decibels.

(11) The terminal isolation provided each subscriber shall be not less than 18 decibels, but in any event, shall be sufficient to prevent reflections caused by open-circuited or short-circuited subscriber terminals from producing visible picture impairments at any other subscriber terminal.

(12) Radiation from a cable television system shall be limited as follows:

Frequencies	Radiation limit (microvolts/ meter)	Distance (feet)
Up to and including 54 MHz...	15	100
Over 54 up to and including 216 MHz.	20	10
Over 216 MHz....................	15	100

(b) Cable television systems distributing signals by using multiple cable techniques or specialized receiving devices, and which, because of their basic design, cannot comply with one or more of the technical standards set forth in paragraph (a) of this section, may be permitted to operate provided that an adequate showing is made which establishes that the public interest is benefited. In such instances the Commission may prescribe special technical requirements to ensure that subscribers to such systems are provided with a good quality of service.

(c) Paragraph (a)(12) of this section shall become effective March 31, 1972. All other provisions of this section shall become effective in accordance with the following schedule:

	Effective date
Cable television systems in operation prior to March 31, 1972_____	Mar. 31, 1977
Cable television systems commencing operations on or after March 31, 1972_____	Mar. 31, 1972

§ 76.609 Measurements.

(a) Measurements made to demonstrate conformity with the performance requirements set forth in §§ 76.701 and 76.605 shall be made under conditions which reflect system performance during normal operations, including the effect of any microwave relay operated in the Cable Television Relay (CAR) Service intervening between pickup antenna and the cable distribution network. Amplifiers shall be operated at normal gains, either by the insertion of appropriate signals or by manual adjustment. Special signals inserted in a cable television channel for measurement purposes should be operated at levels approximating those used for normal operation. Pilot tones, auxiliary or substitute signals, and nontelevision signals normally carried on the cable television system should be operated at normal levels to the extent possible. Some exemplary, but not mandatory, measurement procedures are set forth in this section.

(b) When it may be necessary to remove the television signal normally carried on a cable television channel in order to facilitate a performance measurement, it will be permissible to disconnect the antenna which serves the channel under measurement and to substitute therefor a matching resistance termination. Other antennas and inputs should remain connected and normal signal levels should be maintained on other channels.

(c) As may be necessary to ensure satisfactory service to a subscriber, the Commission may require additional tests to demonstrate system performance or may specify the use of different test procedures.

(d) The frequency response of a cable television channel may be determined by one of the following methods, as appropriate:

(1) By using a swept frequency or a manually variable signal generator at the sending end and a calibrated attenuator and frequency-selective voltmeter at the subscriber terminal; or

(2) By using a multiburst generator and modulator at the sending end and a demodulator and oscilloscope display at the subscriber terminal.

(e) System noise may be measured using a frequency-selective voltmeter (field strength meter) which has been suitably calibrated to indicate rms noise or average power level and which has a known bandwidth. With the system operating at normal level and with a properly matched resistive termination substituted for the antenna, noise power indications at the subscriber terminal are taken in successive increments of frequency equal to the bandwidth of the frequency-selective voltmeter, summing the power indications to obtain the total noise power present over a 4 MHz band centered within the cable television channel. If it is established that the noise level is constant within this bandwidth, a single measurement may be taken which is corrected by an appropriate factor representing the ratio of 4 MHz to the noise bandwidth of the frequency-selective voltmeter. If an amplifier is inserted between the frequency-selective voltmeter and the subscriber terminal in order to facilitate this measurement, it should have a bandwidth of at least 4 MHz and appropriate corrections must be made to account for its gain and noise figure. Alternatively, measurements made in accordance with the NCTA standard on noise measurement (NCTA Standard 005-0669) may be employed.

(f) The amplitude of discrete frequency interfering signals within a cable television channel may be determined with either a spectrum analyzer or with a frequency-selective voltmeter (field strength meter), which instruments have been calibrated for adequate accuracy. If calibration accuracy is in doubt, measurements may be referenced to a calibrated signal generator, or a calibrated variable attenuator, substituted at the point of measurement. If an amplifier is

used between the subscriber terminal and the measuring instrument, appropriate corrections must be made to account for its gain.

(g) The terminal isolation between any two terminals in the system may be measured by applying a signal of known amplitude to one and measuring the amplitude of that signal at the other terminal. The frequency of the signal should be close to the midfrequency of the channel being tested.

(h) Measurements to determine the field strength of radio frequency energy radiated by cable television systems shall be made in accordance with standard engineering procedures. Measurements made on frequencies above 25 MHz shall include the following:

(1) A field strength meter of adequate accuracy using a horizontal dipole antenna shall be employed.

(2) Field strength shall be expressed in terms of the rms value of synchronizing peak for each cable television channel for which radiation can be measured.

(3) The dipole antenna shall be placed 10 feet above the ground and positioned directly below the system components. Where such placement results in a separation of less than 10 feet between the center of the dipole antenna and the system components, the dipole shall be repositioned to provide a separation of 10 feet.

(4) The horizontal dipole antenna shall be rotated about a vertical axis and the maximum meter reading shall be used.

(5) Measurements shall be made where other conductors are 10 or more feet away from the measuring antenna.

§ 76.613 Interference from a cable television system.

In the event that the operation of a cable television system causes harmful interference to reception of authorized radio stations, the operation of the system shall immediately take whatever steps are necessary to remedy the interference.

§ 76.617 Responsibility for receiver-generated interference.

Interference generated by a radio or television receiver shall be the responsibility of the receiver operator in accordance with the provisions of Part 15, Subpart C, of this chapter: *Provided, however,* That the operator of a cable television system to which the receiver is connected shall be responsible for the suppression of receiver-generated interference that is distributed by the system when the interfering signals are introduced into the system at the receiver.

Appendix C

SAMPLE FRANCHISE PROVISIONS ON TECHNICAL STANDARDS AND PERFORMANCE MONITORING

EXTRACT FROM NEW YORK CITY—TelePrompTer FRANCHISE

Section 4(k). The Company shall furnish to its subscribers and customers for all services the best possible signals available under the circumstances existing at the time, to the satisfaction of the Director of Communications, and shall provide quality reception of its Basic Service to each subscriber so that both sound and picture are produced free from visible and audible distortion and ghost images on standard television receivers in good repair.

EXTRACT FROM REDLANDS, CALIFORNIA FRANCHISE

Section 17. OPERATIONAL STANDARDS. The CATV system shall be installed and maintained in accordance with the highest and best accepted standards of the industry to the effect that subscribers shall receive the highest possible service. In determining the satisfactory extent of such standards, the following, among others, shall be considered:

A. That the system provide a minimum of twenty (20) channels, capable of delivering to subscribers the entire VHF and FM spectrum and selected portions of the UHF spectrum.

B. That the system, as installed, be capable of passing standard color TV signals without the introduction of material degradation on color fidelity and intelligence.

C. That the system and all equipment be designed and rated for 24-hour per day continuous operation.

D. That the system provide a nominal signal level of 2000 microvolts across 75 ohms at the input terminals of each TV receiver. A minimum signal of 1000 microvolts across 75 ohms shall be maintained for at least 95 percent of the operating time.

E. The system signal-to-noise ratio shall not be less than 43 db (decibels).

Signal-to-noise is a figure of merit, thus insuring distribution of picture without noticeable degradation.

F. Hum modulation of a 100-percent modulated picture signal shall not exceed 2 percent.

G. The system shall not exceed a VSWR (Voltage Standing Wave Ratio) of 1.2 at any point in the system.

H. The sound carrier level on each television channel distributed shall not be less than 17 db below the level of either adjacent picture carrier.

I. The carrier level of each FM channel distributed shall be not less than 17 db below the picture carrier level in television channels adjacent to the FM band.

J. Co-channel interference, adjacent channel interference and other extraneous signals, including hum, measured at the receiver input, shall be at least minus 40 db with respect to the peak carrier level of each desired channel.

K. Isolation between any two subscribers shall be at least 35 db.

L. Radiation from coaxial cables and electronic equipment in the distribution system, including power supplies and associated power lines, shall be less than 10 microvolts per meter at any point at a distance of 10 feet. Each power supply or its associated housing shall be equipped with a suitable RF power line filter which shall provide not less than 36 db of attenuation to all frequencies transmitting over this system.

M. Interference from sources external to the system shall not be noticeable with a blank-screen test.

N. All equipment must conform with any pertinent City of Redlands and/or underwriter's laboratory standards, whichever shall be more stringent.

O. The system shall maintain all of the above specifications between ambient temperature limits from minus 20 degrees Fahrenheit to plus 120 degrees Fahrenheit, where ambient temperature is defined as the officially recorded City air temperature, and also for AC power line variations between 105 and 130 volts, and between 55 and 65 hertz. The maximum level change at the input to any randomly selected receiver shall be 3 db with a temperature change of 50 degrees Fahrenheit, with no adjustment and with the AC line voltage constant within 5 percent of nominal. The maximum level change shall be 3 db for a constant ambient temperature with a voltage variation from 105 to 130 volts.

P. The signal received by the customer shall be substantially the same quality as originally transmitted by the staion being received taking into consideration the technical conditions over which the system operator has no control.

Section 18. PROOF OF PERFORMANCE SPECIFICATION. The franchise grantee shall submit at the commencement of operation a proof of performance for each CATV system or major operating portion thereof that the system is operating in conformance with each of the standards and specifications listed above as of the date of the statement. The statement shall be submitted in a form approved by the City Manager. Thereafter, at least once annually, the City shall employ the services of a consultant expert in the field, who will determine that the system is operating in conformance with each of the standards and specifications listed above as of the date of the statement. This statement, too, shall be submitted in the form approved by the City Manager. Any such costs incurred by the City to determine proof of performance of the operational specifications shall be paid by the grantee within 10 days of receipt of statement of costs sent by the City. More frequent proof of performance of the operational specifications may be required.

Appendix D

SYSTEM RELIABILITY AND MAINTAINABILITY

Reliability can be defined as the capability of a system or item of equipment to maintain a consistent level of performance, once that performance has been initially established. The less reliable a system is, the more its day-to-day performance will deviate from the norm. This deviation may be either in the form of a malfunction that makes the system (or a portion of it) completely inoperative, or a degradation in performance where the system continues to operate, but not as well as it should.

At least two methods of quantifying and specifying reliability are in general use:

1. A system "up time" (or "down time") can be specified within which the system must continue to operate without malfunction for some percentage of the total operating cycle. For a cable system, as an example, an up time of 95 percent might be specified. This would mean that if the system were designed to operate 24 hours a day, it would have to be in actual service 95 percent of those 24 hours each day (or week, month, etc.) to meet its reliability specification. (Failures or malfunctions that constitute a loss in service also would have to be specified.)

2. A "mean-time-between-failure" (MTBF) can be specified. This might, as an example, require the system to operate satisfactorily on an average of 500 hours between malfunctions. Over the course of perhaps 6 months or a year, the total number of failures would be divided into the total operating time to obtain the actual MTBF figure, which would be compared against the specified figure. Again, terms such as "failure" and "malfunction" require precise definition.

Maintainability is defined as the ease with which a malfunction can be repaired, once it does occur (i.e., how fast the system can be made operable again). Since some failures will occur over the course of time no matter how carefully the system is designed, it is important that the down time can be held to a minimum.

A common quantification of maintainability is the "mean-time-to-repair" (MTTR) figure, which is the average time to repair malfunctions. In an operating system, this is measured by dividing the total time the system is inoperative (again over perhaps 6 months or a year) by the number of malfunctions. As an example, one malfunction in a year that kept the system inoperative for 100 hours would give an MTTR of 100 hr, while 50 malfunctions that resulted in a total down time of 100 hours would mean an MTTR of 2 hours.

Appendix E

FCC DEFINITION OF A CABLE TELEVISION SYSTEM [1]

76.5 Definitions.

(a) *Cable television system (or CATV system).*
Any facility that, in whole or in part, receives directly, or indirectly over the air, and amplifies or otherwise modifies the signals transmitting programs broadcast by one or more television or radio stations and distributes such signals by wire or cable to subscribing members of the public who pay for such service, but such term shall not include (1) any such facility that serves fewer than 50 subscribers, or (2) any such facility that serves only the residents of one or more apartment dwellings under common ownership, control or management, and commercial establishments located on the premises of such an apartment house.

[1] FCC, *Cable Television Report and Order,* 37 Fed. Reg. 3252 (1972).

REFERENCES

1. FCC, *Cable Television Report and Order,* 37 Fed. Reg., No. 30, February 12, 1972, Part 76, Subpart K, Sec. 76.605.
2. John E. Ward, *Present and Probable CATV/Broadband-Communication Technology,* Report ESL-R-449, Electronic Systems Laboratory, Massachusetts Institute of Technolgy, June 7, 1971; prepared for the Sloan Commission on Cable Communications. The main part of the report is reproduced as Appendix A to *On the Cable: The Television of Abundance,* Report of the Sloan Commission on Cable Communications, McGraw-Hill Book Company, New York, c. 1971.
3. W. S. Baer, *Interactive Television: Prospects for Two-Way Services on Cable,* The Rand Corporation, R-888-MF, November 1971.
4. J. E. Ward, *What Belongs on the Cable?,* paper presented at the 21st Annual Convention, National Cable Television Association, May 1972.
5. *Broadcasting,* June 12, 1972, p. 5.

Chapter 4

Citizen Participation in Planning

Robert K. Yin

I. INTRODUCTION

PARTICIPATION: THE CITIZEN'S VIEW

Our aim is merely to conduct a technical feasibility study of cable television in your city. Citizen involvement with cable television might best await the results of the study, which will outline the major technical and economic issues and options, and which can thus form the background for an intelligent community discussion.

—Anonymous cable consultant, at the beginning
of an urban cable-TV feasibility study.

Urban citizens in many cities have already become aware of the potential growth of cable television into American cities and the major urban markets. City after city will probably witness a similar sequence of events: An initial feasibility study will be followed first by the development of tentative franchise terms and public hearings, and then by the awarding of a franchise and construction of a cable system.[1] This sequence of events may be highly publicized, or confined to a relatively intimate group of people; it may take years from beginning to end, or it may occur very quickly. In any event, urban citizens will expect to have a meaningful role in the development of cable television in their community; for telecommunications, especially of the sort made possible by advanced cable television systems, may eventually revolutionize human communication patterns, perhaps as dramatically as did the telephone and the automobile. Consequently, citizens will want to deliberate, discuss, and decide carefully before allowing cable systems to be developed in their cities.

As a first step, during the initial planning period, urban citizens *will not accept* the passive role implied by the consultant's words quoted above. They may even feel unintimidated by the expertise held by study groups, franchise applicants, or other recognized cable "experts." As residents have learned through their experiences in other large-scale public programs, active citizen participation in the planning process is a necessary prelude to any subsequent citizen responsibility in helping to design, operate, or use cable systems. Moreover, citizen participation that begins only with the public hearings on cable franchises may be too late to be effective.[2]

---

[1] The decisionmaking process is described in Walter S. Baer, *Cable Television: A Handbook for Decisionmaking*, Crane, Russak & Co., New York, 1974, and in Chapter 5 of this book.

[2] See the remarks of Stuart Sucherman in Charles Tate (ed.), *Cable Television in the Cities: Community Control, Public Access, and Minority Ownership*, The Urban Institute, Washington, D.C., 1971, p. 79.

By that time, many options already may have been eliminated. Instead, the public will want to make its desires known during the initial study period, when it can shape the very questions that are to be asked, and not merely offer alternate answer to narrowly conceived questions.

PARTICIPATION: THE MUNICIPAL OFFICIAL'S VIEW

For municipal officials, citizen participation in the planning process can raise many delicate issues. While participation is highly desirable in principle, it often leads to either of two less desired outcomes. First, it may amass such a divergence of citizens' views that no public consensus can be reached—an outcome easily imagined, since today's cities are composed of many diverse interest and neighborhood groups. The groups may disagree, for instance, over the appropriate forms of ownership, and whether a regional system should be owned by a single operator, or whether regional coverage can be accomplished through separate ownership of the individual hubs that make up a regional network. The municipal official, then, must be prepared to deal with diverging opinions and arbitrate among competing groups, and this role may create extreme political hostility.

Second, citizen participation may jeopardize the building of any cable system. In the past, citizens have often shown their power by preventing or delaying the construction of new buildings and highways. As for cable television, participation could lead to consensus around unrealistic alternatives, such as a demand that control over all revenues be given to the community, in which case few cable operators will even want to bid for a franchise. Whether obstructionist positions are justifiable or not, the municipal official must weigh the advantages of citizen participation against the risk that it will thwart the development of a cable system in his community.

If all goes well, however, citizen participation can yield many beneficial results. For instance,

- *Successful participation could increase interest in cable television, and potentially enlarge the subscriber base.*

Urban cable television systems will rely almost exclusively on individual subscribers for financial support. A positive experience by a large number of citizens during the planning process will increase the likelihood that people will subscribe to and use the cable system. Moreover, any changes in the cable-system design due to citizen participation presumably render the system more relevant to citizens' needs, and hence should increase the potential subscriber base.

- *Successful participation could result in improved public service use of cable television.*

In addition to new entertainment, the new urban cable systems promise to provide citizen access to television programming and such new public services as fire and burglar alarm systems, new health and social services, at-home adult education, and neighborhood-specific services.[3] These services will be more client-oriented, and hence perhaps more effectively delivered, if the citizens who are their potential users have participated in the design and planning stages.

- *Successful participation could increase the communication effectiveness of cable television.*

Cable television brings two distinctly new communication capabilities. The first is the ability to pinpoint highly specific audiences, whether by geographic area or by any other criterion. In other words, specific groups of people will be able to use cable TV to communicate with each other. The second is the ability to provide two-way data, audio and video interaction. Although two-way cable services will not be widespread for at least several years, citizens could encourage the early use of these capabilities; they are best able to define their own communication patterns, and the better the cable design suits these patterns, the more likely the cable system will be used.

- *Citizen participation can enable blacks and other inner-city minority groups to influence the development of cable systems.*

Because American cities are increasingly composed of blacks and other minority groups, the new urban cable television must accommodate to their interests. In some cases, these groups are concerned only with having access to cable systems and creating programs of local interest. In other cases, they are also interested in bidding for the control and ownership of the cable system.

Municipal officials will thus be weighing the costs of broad citizen participation (e.g., political divisiveness and delay in developing a cable system) against the benefits (e.g., increasing the potential subscriber base, improving the design of the cable system, and allowing minorities the opportunity to own and control cable systems). In addition, officials should be aware that the private cable operators who may bid on the cable franchises may also have begun their own negotiations with citizens' groups. The operators know they will have to deal with citizens, if only as subscribers, and that it behooves them to cultivate good will and cooperation at the very beginning.

* * * * * * * *

Each official will decide about citizen participation in cable television on the basis of the unique circumstances in his own city. Some officials may decide to minimize citizen participation to accelerate the franchising process, but will do so at the risk of having few bidders, a system poorly designed to serve community needs, or a low penetration rate. For those who decide to attempt some form of citizen participation, many issues still need to be resolved. The purpose of this report is to provide guidelines for engaging citizen participation in the cable planning process. The report discusses some of the forms participation can take, some of the benefits that can be expected, and some of the critical issues

---

[3] For a discussion of public access and municipal services, see the companion reports by Richard Kletter, "Making Public Access Effective," and Robert K. Yin, "Applications for Municipal Services," Chapters 1 and 2, respectively, of Polly Carpenter et al., *Cable Television: Developing Community Services*, Crane, Russak & Co., 1974.

and problems. One topic not covered here is citizen participation in the process by which a cable system receives a certificate of compliance from the Federal Communications Commission after a franchise has been awarded. Guidelines for participation in the certification process and other legal proceedings are covered in Chapter 5 of this book.

One point should be made clear at the outset: no external document, including this one, can provide a definitive set of guidelines for citizen participation, for the very spirit of such participation is that specific people in a specific locale at a specific time will decide how to organize both the *form* and the *content* of their participation. To appreciate this point fully, and to understand better the contemporary urban milieu, we first turn to a brief review of how citizen participation has developed on the urban scene.

II. SETTING THE SCENE

Citizen participation has a considerable history in the American city. The first two parts of this section describe local and federal developments bearing on citizen participation over the last several decades. The third part attempts to show the relationship between these developments and the entrance of cable television into the contemporary city, by explicitly pointing to the lessons for cable television. (The time-constrained reader may want to skip the first two parts and turn immediately to "Lessons for Cable Television," below.)

EARLY DEVELOPMENTS: REFORM TO RENEWAL

The reform movement in city government began late in the nineteenth century, but exerted its greatest force during the first half of this century. The movement did not emanate from a coherent plan to change city government, but from the spontaneous reactions of many citizens to the excesses of the traditional spoils system and the political machines then characteristic of city government.[4] A major reform was to have merit-based civil service systems gradually replace patronage politics in the selection and retention of public employees (Congress passed the Pendleton Civil Service Act in 1883). It is more relevant to our present subject, however, that reform was accompanied in most cities by expanding participation of private citizens in municipal affairs. As part of the full elaboration of mugwump politics and reform, citizens formed civic associations, citizens' leagues, city clubs, research bureaus, and other citizen-supported organizations to analyze and oversee the affairs of government. These citizens were usually highly educated and well-to-do. Although their organizations seldom wielded any real power or influence, their activities at least established formal channels of government-citizen interaction on community needs and issues.[5]

A second brand of citizen participation developed through numerous service and neighborhood organizations, such as settlement houses, block associations, self-

---

[4] See Richard Hofstadter, *Age of Reform,* Knopf, New York, 1955, esp. pp. 257-271.

[5] For descriptions and partial evaluations of the citizens' organizations, see ibid.; Lorin Peterson, *The Day of the Mugwump,* Random House, New York, 1961; Edward C. Banfield and James Q. Wilson, *City Politics,* Harvard University Press, Cambridge, Massachusetts, 1963, pp. 142, 250-258; and Daniel Bell and Virginia Held, "The Community Revolution," *The Public Interest,* No. 16, Summer 1969, pp. 142-177.

help groups, and religious and ethnic groups.[6] These organizations typically dealt with local "social problems" and the immediate conditions of the neighborhood, involving a more grass-roots level of support than did the reform movement. Citizen participation ranged from support for the more benign forms of social work to the organization of militant protest groups. Among the protest groups, perhaps the best known were those organized by Saul Alinsky. Alinsky worked on his first successful project in the late 1930's, when he helped laborers and neighborhood residents to gain concessions from city hall to improve life in Chicago's depressed stockyards area.[7] Just as the reformers had shown how citizens could participate in government through formal channels, these neighborhood-based organizations established useful precedents for citizens to challenge governmental action or inaction at the street level.

A third occasion for citizen participation paralleled the rise of the city planning profession. The planners attempted to bring a more orderly approach to the haphazard development of urban services and residential patterns that accompanied city growth. By the 1920's, city planners, mostly trained initially as civil engineers or architects, had helped to establish planning and zoning as municipal responsibilities.[8] At first, the function of the planners appeared to complement the reform movement, for planners, civic leaders, and fusion reform candidates were natural allies against the political machines. In Philadelphia, for instance, supporters of a losing fusion candidate in 1939 created a city policy committee that worked for the development of a planning commission; eventually, a regional planning organization was founded, involving both the planning commission and citizens' councils.[9] But the major accomplishment of city planning was to bring new methods of land-use and facilities-planning to the city. These were supposed to result in more rational use of the city's physical structure. Perhaps most characteristic of the planners' approach was the development of uniform standards for the quantity and quality of municipal facilities. For instance, as early as 1890, standards were being developed for the number, size, and equipping of playgrounds; similar methods were later applied to hospitals, libraries, parks, schools, and even neighborhoods.[10]

The creation of plans required a certain amount of interaction between local government and citizen groups. Though the comprehensive plan covered a large area, its building block was usually the neighborhood unit, which had a specified size

[6] These are typically the organizations created by professional social workers. For general descriptions of such organizations, see Mayer N. Zald, "Sociology and Community Organization Practice," in Mayer N. Zald (ed.), *Organizing for Community Welfare*, Quadrangle, Chicago, 1967, pp. 27-61; Ralph M. Kramer and Harry Specht (eds.), *Readings in Community Organization Practice*, Prentice-Hall, Englewood Cliffs, New Jersey, 1969; and Fred M. Cox, et al. (eds.), *Strategies of Community Organization*, Peacock, Itasca, Illinois, 1970.

[7] Alinsky has applied similar tactics and strategies elsewhere over the last thirty years. See Saul D. Alinsky, *Reveille for Radicals*, University of Chicago Press, Chicago, 1946; idem, "The Professional Radical, 1970," *Harper's Magazine*, January 1970, pp. 35-42; and idem, *Rules for Radicals*, Random House, New York, 1971. For contrasting views of his work, see Daniel P. Moynihan, *Maximum Feasible Misunderstanding: Community Action in the War on Poverty*, Free Press paperback edition, New York, 1970, p. 185 (complimentary); and Frank Riessman, "The Myth of Saul Alinsky," in *Strategies Against Poverty*, Random House, New York, 1969, Chap. 1 (critical).

[8] For a brief history, see Herbert J. Gans, "City Planning in America: A Sociological Analysis," in *People and Plans: Essays on Urban Problems and Solutions*, Basic Books, New York, 1968, pp. 57-77.

[9] John W. Bodine, "The Indispensable One-Hundredth of 1 Percent," in H. Wentworth Eldredge (ed.), *Taming Megalopolis*, Doubleday Anchor, Garden City, New York, 1967, pp. 956-971.

[10] Gans, op. cit., p. 59.

and a well-defined set of public facilities.[11] The neighborhood unit could make the plan directly relevant to the interests of various neighborhood groups. Whatever the exact nature of the plan, however, the citizens' role was mainly passive: reviewing and ratifying the planners' decisions.[12] The planners, sometimes after years of study, would present near-final plans intimately affecting a neighborhood's environment through zoning changes or the selective placement of municipal facilities; the citizens would review such plans at public hearings, but often had no resources and little time to do more than acquiesce to the planners' preconceived notions, which seemed to offer a varied set of options but nevertheless perpetuated a narrow value system.

Furthermore, because of their professional training, the traditional planners were so absorbed with the physical environment of the city that they tended to disregard its social aspects. This adherence to physical determinism[13] made them insensitive, at least in appearance, to those very aspects of life that residents considered most important. For the citizen whose exposure to the planning process may have been his only direct encounter with municipal government, this experience created a sentiment that associated government with buildings, highways, physical facilities, programs, and generally everything except people and human life. The planning profession itself has subsequently reacted by giving greater emphasis to social science and the "real world" in its work.[14] Moreover, a new type of "advocacy planning," in which planners deliberately work on behalf of neighborhood residents to protect community interests in dealing with government, became popular in the 1960's.[15] But these changes have not yet been able to undo the antipopulist sentiment linked with planners and local government.

Aside from the planners' traditional orientation toward the physical environment, two specific aspects of their work also colored the early citizen-planner interactions. One aspect was the increasing tendency for planners to design citywide master plans and regional plans for metropolitan areas. The broad-scale planning suggested a disregard for local issues, particularly within the central city. Another aspect was the planners' role in the federal government's early urban development programs, i.e., public housing and urban renewal. Public housing projects had been initiated under the National Industrial Recovery Act of 1933, but the major urban renewal and public housing funds were provided by the Housing Acts of 1949 and 1954. The Housing Act of 1954 also embodied specific provisions for citizen participation. In these programs, local citizens mainly performed advisory functions, taking

[11] Ibid., p. 64. See also Clarence A. Perry, "The Neighborhood Unit," Monograph I of Vol. 7, *Regional Survey of New York and Its Environs,* Committee on Regional Planning, New York, 1929, pp. 22-140; and Lewis Mumford, *The Urban Prospect,* Harcourt, Brace & World, New York, 1956, pp. 56-78.

[12] David R. Godschalk, "The Circle of Urban Participation," in H. Wentworth Eldredge (ed.), op. cit., pp. 971-979. For other descriptions of the planner-citizen relationship, see Nathan Glazer, "The School as an Instrument in Planning," *J. Amer. Inst. Planners,* Vol. 25, November 1959, pp. 191-196; and James Q. Wilson, "Planning and Politics: Citizen Participation in Urban Renewal," in James Q. Wilson (ed.), *Urban Renewal: The Record and the Controversy,* M.I.T. Press, Cambridge, Massachusetts, 1966, pp. 407-421.

[13] In its most simplified form, physical determinism assumes that the physical environment determines the behavior and social relationships of the residents. See Gans, op. cit., p. 64.

[14] See ibid.; Jane Jacobs, *The Death and Life of Great American Cities,* Random House, New York, 1961; Leonard J. Duhl (ed.), *The Urban Condition: People and Policy in the Metropolis,* Basic Books, New York, 1963; and Ernest Erber (ed.), *Urban Planning in Transition,* Grossman, New York, 1970.

[15] Paul Davidoff, "Advocacy and Pluralism in Planning," *J. Amer. Inst. Planners,* Vol. 31, November 1965, pp. 4-12.

the form of middle-class, blue-ribbon committees that had only a marginal impact on the programs designed.[16] The few cases of successful participation and renewal were limited to economically mixed neighborhoods, such as the Hyde Park-Kenwood area in Chicago;[17] in cases where entirely lower-class neighborhoods were involved, no really adequate procedures for accommodating both participation and renewal were ever adopted, and citizen participation and urban renewal came to occupy antagonistic positions: the more of one, the less likely the other.[18] The general effect of these programs was thus to associate planners and local government with the displacement of people from their original homes and neighborhoods, with little attention paid to the relocation problems of these people.[19] If "slum clearance" was not merely a term for the clearance of blacks and other minority groups, it was certainly a substitute label for "people clearance."

Important and still relevant features of citizen participation thus developed out of these early experiences, whether participation took the form of reform-oriented civic associations, neighborhood and protest groups, or citizens dealing with the city planning process. Though citizen participation had not yet led to issues of community control over actual resource allocation, it had nevertheless become clearly identified with grass-roots democracy. Distinct channels of access now existed for citizens to attempt to influence directly the changes occurring in their cities. Moreover, on any given issue citizen participation had by this time also become clearly associated with matters involving people's lives, their homes, and the social quality of the urban environment. Physical environment, new technology, official bureaucracies, and large-scale programs tended to be defined as adversaries serving special interests, but not serving the local people. On the other hand, citizen participation was still mostly a white, middle-class affair; the integration of the urban poor and minority groups into the participation process was still to come.

MIDCENTURY: THE FEDS TAKE OVER

Urban renewal and public housing were only the first of several federal programs directly affecting the city and involving citizen participation. But whereas citizen participation was rather perfunctory in the renewal program, by the end of the 1960's participation was to assume a much different role. The changes occurred rapidly, with citizens involved in the slum renewal and youth-delinquent projects

[16] Stephen D. Mittenthal and Hans B. C. Spiegel, "Urban Confrontation: City Versus Neighborhood in the Model Cities Planning Process," Columbia University, New York, 1970, p. 31.

[17] For coverage of the events in Chicago, see Julia Abrahamson, *A Neighborhood Finds Itself,* Harper, New York, 1959; and Peter H. Rossi and Robert A. Dentler, *The Politics of Urban Renewal: The Chicago Findings,* Free Press, Glencoe, Illinois, 1961. For events in other cities, see Langley Carleton Keyes, Jr., *The Rehabilitation Planning Game: A Study in Diversity of Neighborhood,* M.I.T. Press, Cambridge, Massachusetts, 1969; and J. Clarence Davies III, *Neighborhood Groups and Urban Renewal,* Columbia University Press, New York, 1966.

[18] Wilson, "Planning and Politics," pp. 416-417.

[19] The classic description of residential life before displacement is Herbert J. Gans, *The Urban Villagers,* Free Press, New York, 1963; see also Marc Fried, "Grieving for a Lost Home," in Duhl (ed.), *The Urban Condition,* pp. 151-171.

that were forerunners of the Anti-Poverty program,[20] and later with the Anti-Poverty program itself and the Model Cities program. With each new program, the federal government increased its support of the participation process and involvement with local affairs. Today, several major federal agencies sponsor projects that call for some type of citizen participation (e.g., the Office of Economic Opportunity, the Department of Housing and Urban Development, the Department of Health, Education, and Welfare, and the Department of Labor),[21] and the federal role has become so dominant that one observer has audaciously suggested that, "When historians examine the events of the 1960s with the hindsight of another generation, it is possible that they will describe the development of *federal standards for citizen participation* as the decade's most significant innovation in the practice of public administration."[22]

The federal program with the greatest influence on citizen participation was the Anti-Poverty program established by the Economic Opportunity Act of 1964. This act called for the creation of Community Action Programs in many cities. The best-known part of the act stated that these programs were to be "developed, conducted, and administered with the maximum feasible participation of residents of the areas and members of the groups served." As several analysts have subsequently pointed out, the origins of this landmark clause are remarkably difficult to trace, as it did not stem directly from prior federal programs or from a strong lobbying effort by any professional, political, or citizen group.[23] There is little evidence of discussions of the clause, and it was apparently inserted at some point in the drafting of the legislation.[24] One guess is that the clause emanated from the generally pluralistic ethos that had been created by the Civil Rights Act of 1964, and that maximum feasible participation was a by-product of the political climate of the time.

As with its origins, the meaning of "maximum feasible participation" was also unclear. As a result, citizen participation took on a wide variety of forms, depending on the specific local circumstances.[25] However, with residents seizing the initiative, participation eventually became linked with two types of activities: (1) residents of inner-city neighborhoods, mostly poor people, were employed in subprofessional jobs associated with the administration of the Anti-Poverty programs, and (2) residents

[20] These were largely supported by the Juvenile Delinquency and Youth Offenses Control Act of 1961 and the Gray Areas programs of the Ford Foundation. The best-known of the youth-delinquent projects was Mobilization for Youth. Planning for this organization began in 1957, with major financial support starting in 1959, and operating services commencing in 1962. For a description of this and other pre-Anti-Poverty programs, see Peter Marris and Martin Rein, *Dilemmas of Social Reform: Poverty and Community Action in the United States,* Atherton, New York, 1967; Harold H. Weissman (ed.), *Community Development in the Mobilization for Youth Experience,* Association Press, New York, 1969; and Moynihan, op. cit.

[21] For an illustrative review of how the different federal programs can operate in one local area, see Melvin B. Mogulof, "Citizen Participation: The Local Perspective," The Urban Institute, Washington, D.C., March 1970.

[22] David M. Fox, "Federal Standards and Regulations for Participation," in Edgar S. Cahn and Barry A. Passett (eds.), *Citizen Participation: Effecting Community Change,* Praeger, New York, 1971, p. 130.

[23] Lillian B. Rubin, "Maximum Feasible Participation: The Origins, Implications, and Present Status," *The Annals,* Vol. 385, September 1969, pp. 14-29.

[24] The legislative history is traced briefly in Ralph M. Kramer, *Participation of the Poor: Comparative Community Case Studies in the War on Poverty,* Prentice-Hall, Englewood Cliffs, New Jersey, 1969, pp. 8-9. See also Marris and Rein, op. cit.; Moynihan, op. cit.; and Rubin, op. cit.

[25] See Dale Rogers Marshall, "Public Participation and the Politics of Poverty," in Peter Orleans and W. R. Ellis, Jr. (eds.), *Race, Change, and Urban Society,* Sage, Beverly Hills, 1971, p. 458; and Mogulof, op. cit., March 1970.

were also engaged in policymaking, program design, and implementation. The latter role was fulfilled through membership on the local governing boards that administered the Anti-Poverty funds.

These participatory activities opened the way for new social and political relationships. Inner-city residents were now officially mandated to speak and act on their own behalf as well as to design and implement programs to meet their own felt needs. This policy recognized that poor people, particularly in contrast to outside professionals, brought some measure of expertise to the scene. As one observer has remarked, it meant support for the idea that "... those who have lived with a problem are in the best position to recognize and understand it. They can 'tell it like it really is.'"[26] Professionals in all fields of endeavor, long accustomed to their preeminence in program development, now had to contend with a new group of "experts" who challenged the very relevance of any outsiders' experiences. On the political side, the new type of participation meant the exercising of real political power, in which neighborhood groups, using federal resources, threatened to establish a new hierarchy of power completely parallel to the elected hierarchy of municipal government.[27]

Partly because of the implications of these new relationships, the Anti-Poverty program was continually challenged by local politicians, and the program eventually had to yield to these political pressures. The federal government turned to a new program, the Model Cities program, enacted by the Demonstration Cities and Metropolitan Development Act of 1966. While still calling for citizen participation, the Model Cities program differed deliberately from the Anti-Poverty program by specifying that program design and implementation had to be approved directly by the local municipal government and operated within its budgetary and personnel systems. In other words, all new activities, including citizen participation, would have to be carried out within the established local political framework, and not outside of it. Nevertheless, community residents once again struggled to extract as much control of the program from the local city hall as possible, but success occurred rarely and only temporarily.[28] At this time, the final outcome of the Model Cities program is unclear, both in terms of the participation issue and the question of whether municipal service delivery has actually improved in the Model Cities neighborhoods.

Federal support and promotion of citizen participation in the Anti-Poverty and Model Cities programs have thus had a dual effect on the urban community. First, the federal government helped to set the precedent for a new type of participation, where control, and not education or advice-giving, is the main issue. The historical trend among programs has been away from merely advisory participation, as in the urban renewal programs, and toward participation as a prelude to decentralized

---

[26] Warner Bloomberg, Jr. and Florence W. Rosenstock, "Who Can Activate the Poor—One Assessment of 'Maximum Feasible Participation,'" in W. Bloomberg and H. J. Schmandt (eds.), *Poverty and Urban Policy,* Sage, Beverly Hills, 1968.

[27] See especially Moynihan, op. cit., pp. xix, 131; and Marshall, op. cit., pp. 454-455.

[28] Much of the early planning experiences in the Model Cities program is traced in Roland L. Warren, "Model Cities First Round: Politics, Planning, and Participation," *J. Amer. Inst. Planners,* Vol. 35, July 1969, pp. 245-252; Mittenthal and Spiegel, op. cit.; and Marshall Kaplan, *The Model Cities Program: The Planning Process in Atlanta, Seattle, and Dayton,* Praeger, New York, 1970. For a brief case history of one city's experience, see Ginger Rosenberg, "Model Cities—Dayton Plays the Game," in Cahn and Passett, op. cit., pp. 271-286.

control, as in the Model Cities programs.[29] Within any given program, the participation of specific citizen groups has also tended to drift in the same direction, that is, toward questions of control over policy and decisionmaking.[30] Second, the federal programs extended participation directly to the urban poor and black people, groups that previously had been excluded from the participation process in the traditional form of citizen participation. This does not necessarily mean that citizen leadership has always been exercised by the poor themselves; in fact, the evidence suggests that the spokesmen for the poor have been middle-class and educated.[31] But it does mean that local residents feel that they have the right—and the right of access—to make their opinions known. In these two senses, a new citizen participation was created in the 1960's.[32]

LESSONS FOR CABLE TELEVISION

These events in citizen participation provide the background against which cable systems will enter the contemporary American city and the top television markets. Since specific cable TV issues will be discussed in the next section, the aim here has been merely to point out that cable TV embraces enough community-oriented aspects to make residents' recent experiences with federal programs and citizen participation highly relevant. In other words, though cable technologists and financiers may view their own entrance into the top television markets merely as part of the tapping of a new and much enlarged subscriber base, urban residents may view the coming of cable television as the latest of a series of incursions into the metropolis by new types of hardware, promoted by powerful (and nonlocal) technological, political, and business interests. Thus, several participatory issues are likely to emerge in most cities that attempt to develop new cable television systems. These include the following:

- *Community residents may view the planning for cable television with deep suspicion.*

This conclusion follows from citizens' previous experiences with participation in public programs: the more experiences, the greater the suspicion. Black residents may be particularly cynical; as one black activist puts it:

There is one discernible thread woven through all of this: It is the fact that the people who determined the structure and administration of all these

[29] The trend is traced in Melvin Mogulof, "Coalition to Adversary: Citizen Participation in Three Federal Programs," *J. Amer. Inst. Planners,* Vol. 35, July 1969, pp. 225-232; idem, "Citizen Participation: A Review and Commentary on Federal Policies and Practices," The Urban Institute, Washington, D.C., January 1970; and Bell and Held, op. cit., pp. 149-153. For typologies of the participation process, see Edmund M. Burke, "Citizen Participation Strategies," *J. Amer. Inst. Planners,* Vol. 34, September 1968, pp. 287-294; and Sherry R. Arnstein, "A Ladder of Citizen Participation," *J. Amer. Inst. Planners,* Vol. 35, July 1969, pp. 216-224.

[30] Mogulof, "Citizen Participation," March 1970.

[31] Neil Gilbert and Joseph W. Eaton, "Who Speaks for the Poor?" *J. Amer. Inst. Planners,* Vol. 36, November 1970, pp. 411-416; and James W. Davis, "Decentralization, Citizen Participation, and Ghetto Health Care," *Amer. Beh. Sci.,* Vol. 15, September-October 1971, pp. 94-107.

[32] The distinction is made especially cogently by Davis, op. cit., pp. 98-99.

programs [public housing, urban renewal, FHA home loans, and anti-poverty] were thinking only of what is good for white people. *The black community determined nothing and controlled nothing ...*

That is why I must speak today as I do. That is why I must tell you that regardless of the individual sincerity of many of you fine people here today, the Model Cities program is looked upon with suspicion by thoughtful black people in Chicago.

Yes, suspicion was and still is part of the climate whenever we hear of any program involving government funds or powerful white institutions.[33]

Suspicions may be especially acute, in both inner city and suburban areas, because of the huge profits people imagine to be realizable from cable systems.[34] Residents may assume that every cable-related action, no matter how beneficent in appearance, really stems from profit-maximizing motives. This applies to research activities as well as business and political activities, because most types of research, whether involving feasibility studies, marketing surveys, demonstration projects, or conferences, can also serve the ulterior profit motive. Suspicion may also be aroused because cable is still so dominantly an issue of technological hardware, and residents may wonder whether the new brand of technologists, like the traditional planners before them, will have the sensitivity and compassion to understand that, in the final analysis, the quality of human social life is at stake.

- *Community residents may expect recognition as experts in their own right on the design and construction of cable systems.*

Since an important feature of cable television is its ability to cater to specialized audiences and to the needs of specific neighborhoods, residents will assume that they have the most (and perhaps the only) expertise in defining local needs. Especially because of their experiences with the Anti-Poverty and Model Cities programs, residents may expect to deal with cable technologists and entrepreneurs on a peer basis, sharing the planning and development of the cable system. Moreover, when residents feel they need advice to supplement their own knowledge, they can and will use their own technical consultants, who include a pool of city planners, lawyers, academicians, and newly emerging experts on cable television itself. This practice follows logically from the methods used by cable technologists: Whenever they have felt the need to understand community issues more perceptively, they have employed their own community "experts."

- *Community residents may demand control over the operation of new cable systems.*

This demand also comes directly from the experiences with federal programs, wherein the historical trend has been toward control, and not passive advice-giving. The trend is not likely to reverse, with the tendencies toward community control and

[33] Arthur M. Brazier, "What Kind of Model Cities?," in Eddie N. Williams (ed.), *Delivery Systems for Model Cities,* University of Chicago Press, Chicago, 1969, pp. 7-13.

[34] For several years, at least, these profits are likely to be more imagined than real. See Baer, op. cit., for a sober view of cable economics.

decentralization stronger than ever. In one sense, the Anti-Poverty and Model Cities programs may be but the beginning of a newer phase of citizen participation, one that will see the development of neighborhood government and decentralized control over such public facilities as health centers, the police, and schools.[35] Thus, the planning for cable television definitely can no longer call for citizen participation in the form of blue-ribbon advisory groups. Local residents would interpret such a call as a failure to recognize the changes that have occurred and the new political ethos that now exists in the cities.[36] At a minimum, residents will be offended by the lack of insight; at a maximum, they may close their doors—collectively and individually—and act to prevent the development of any cable system.

- *Citizen participation can delay or prevent the development of any cable system.*

Not to be forgotten is the ever-present threat that citizen participation can lead to indecision and ultimately to the failure to build any cable system. Such a negative outcome has occurred on other occasions, when citizens have blocked the construction of new housing or new highways and often obstructed federal programs they have participated in. This is not necessarily a poor outcome; it may be proper to delay cable franchising until a local political consensus has been reached. In the past, franchising proceedings have erred on the side of overly hasty, rather than overly protracted, decisionmaking.

Municipal officials should realize, however, that citizen participation may add a new and uncertain factor to the franchising timetable. And if attention is diverted toward either less important or more divisive issues, extensive and unproductive delays can result. The following sections of this report offer guidelines to where citizen participation can be most productive, and what forms participation might take.

[35] On the potential direction of community control and neighborhood government, see Milton Kotler, *Neighborhood Government,* Bobbs-Merrill, Indianapolis, 1969; Alan A. Altshuler, *Community Control: The Black Demand for Participation in Large American Cities,* Pegasus, New York, 1970; and Douglas Yates, "Neighborhood Government," in Robert K. Yin (ed.), *The City in the Seventies,* Peacock, Itasca, Illinois, 1972, pp. 119-127.

[36] Marshall, op. cit., p. 479; and Irving Lazar, "Which Citizens to Participate in What?," in Cahn and Passett, op. cit., p. 107.

III. COMMUNITY ISSUES IN THE PLANNING OF CABLE TELEVISION SYSTEMS

Cable television brings new entertainment and public services directly into a resident's home. In addition, the potential for two-way audio and video interactions will make cable television a far more personalized medium of communication than is broadcast television. Finally, cable television brings a new entrepreneur on the urban scene who will build the cable system, solicit subscription fees, and attempt to make profits, all in the name of "serving the community." Because of these characteristics of cable television, every resident may feel legitimately entitled to be active in the design and operation of cable systems.

As a general guideline, the spirit of citizen participation dictates that any issue deemed important by citizens be open for consideration. Cable television planners and municipal officials may be surprised if they attempt to impose their own agenda on citizen groups, particularly if they try to limit citizen deliberation to a few issues, such as the desirable types of new programs, public access to programming time, and support for locally originated programs. While new programming is an important issue, such a focus can have negative connotations: citizens may feel that their interests are being limited to the creation of a citizens' channel, a black people's channel, or whatever special channel is required, and that they have been eliminated from the more important questions of community ownership, control, and profit.

Actually, a wide range of issues can be covered by citizens, cable planners, and municipal officials during the planning for new cable systems. The following list is intended as a sampling of these issues:

OWNERSHIP

- *Will the cable system be owned by a private entrepreneur, a nonprofit group, or the municipality itself? In any case, what are the possibilities for community control, especially by racial and ethnic minorities?*

The major issue here is that cable television is the newest representative of an industry, television and radio communications, in which black and minority ownership has been conspicuously absent. Currently,

242

Blacks own none of the more than 900 licensed over-the-air television stations. Blacks own only about 17 of the 350 or so "soul" radio stations that cater to black audiences. Blacks are participating in the ownership of only two . . . of the more than 4,500 cable-TV franchises that have been awarded by municipal officials to date.[37]

Since many American cities now have sizable black populations, black leaders across the country feel it only appropriate that new urban cable systems be operated by black community groups. They see cable television as perhaps the last opportunity for black people to gain some control over the mass communications industry and to serve the black audience with programs of interest to black people. Thus, blacks will be exceedingly interested in all phases of cable system development, including design, construction, programming, and channel access.

There are many possible forms of cable ownership. The most common up to now has been private ownership. Most observers expect that private cable operators will continue to dominate the scene, and private ownership is probably the most reliable form for developing and operating a new cable system. In conjunction with citizen groups, however, municipalities may want to explore other arrangements, including the formation of a nonprofit organization, municipal ownership, or some sort of joint venture between community groups and private operators.[38]

CABLE SYSTEM GEOGRAPHY

- *How extensive an area will the cable system cover? How many geographic subsystems, if any, should be designed? What areas should the subsystems cover? How should trunk lines for each subsystem be laid out?*

There are no fixed rules regarding the geographic size of a cable system. Each municipality has the jurisdiction to grant its own cable franchise or to subdivide its area among several cable franchises. Moreover, many municipalities in a region may jointly grant a single franchise or multiple franchises covering the whole region. Such cooperation among municipalities would be needed, for instance, if an entire metropolitan area were to have a unified cable system. The decision on how much territory a cable system should cover could be a part of the citizen participation agenda.

As a general guideline, cable systems usually will benefit from regionalization and wide-scale technical compatibility. This is because much of the service that cable television can provide lies in its ability to support programs directed toward very specific audiences. These audiences may be geographically contiguous, as in a neighborhood, or they may be geographically dispersed, as with a particular interest group, such as musicians or artists. A cable system will be of greater service if it covers a large geographic area, since the wider coverage permits the servicing of both geographic-contiguous and geographic-dispersed audiences. Moreover, regional

---

[37] Monroe W. Karmin, "Blacks Seeking Control of Big-City Cable TV Face Uphill Struggle," *The Wall Street Journal*, December 29, 1971.

[38] For more on ownership arrangements, see Chap. 4 of Baer, op. cit.; on joint ventures, see footnote 42 below.

cable systems, covering entire metropolitan areas and possibly connected via satellite relays with other metropolitan areas, can take advantage of many economies of scale.

However, regional cable systems tend to increase the scale beyond the means for local control. The larger any system, the more likely it will be controlled by distant and powerful institutions. To this extent, the design of regional cable systems may remind residents of the master planning for metropolitan areas and regions, a practice that is usually considered incompatible with the desire for community control and decentralization.[39] Regionalization of cable television may also be interpreted as being similar to the attempt to install metropolitan government. This move has also been associated with the desire to prevent control of the central city from falling into the hands of the inner-city residents, especially because of the rising proportion of black residents in the central city.[40]

One possible compromise between the need for regionalization and the need to assure local control over cable systems is to examine the possibilities for *multiple-hub cable systems*. A multiple-hub system is composed of a number of separate hubs, or cable subsystems, each having a headend at its center and each providing television to a circular area with approximately a 5-mile radius. Each hub can be owned and operated by a different group, but still be technically compatible with the others, so that in its totality the entire multiple-hub system can cover a metropolitan area or even a larger regional area. The multiple-hub cable system can thus provide a choice of different programs for each hub, similar programming for a few but not all of the hubs, or the same region-wide programming for the entire system.

This type of multiple-hub system was recently studied in relation to the cable needs for Dayton, Ohio.[41] As a result of the study, a central city community group and a private cable operator have created a joint venture and plan to bid for the ownership of one of the central city hubs.[42] For other municipalities, citizen participation in the planning stage could help to determine how many separate hubs, if any, the cable system in its community should have and where they should be located.

A second characteristic of cable geography that citizens can discuss is the pattern of trunk lines emanating from each headend. The trunk lines can further differentiate the audience, since residents on different trunk lines can receive different programs. In other words, even within each hub, cable technology can allow for selectivity in directing specific programs to specific audiences. Thus, just as with the number and location of the hubs, citizen discussion of the number and location of trunk lines would seem appropriate, to help identify distinct neighborhoods within a hub that should have their own distinct programming.

[39] Banfield and Wilson, op. cit., pp. 187-203.

[40] See, for example, Frances Fox Piven and Richard A. Cloward, "Black Control of Cities: Heading It Off by Metropolitan Government," *The New Republic*, September 30 and October 7, 1967.

[41] The system is described in detail in Nathaniel E. Feldman, "System Designs for the Dayton Metropolitan Area," in L. L. Johnson et al., *Cable Communications in the Dayton Miami Valley: Basic Report*, The Rand Corporation, R-943-KF/FF, January 1972, Paper 1.

[42] See the prospectus of this joint venture, "Plan for Minority Participation in a Cypress Cable System in Dayton, Ohio," Cypress Communications Corporation, February 23, 1972. The venture is called "Cypress Southwest," a new corporation equally owned by the Cypress Communications Corporation and the Citizens Cable Corporation. The new corporation will bid for ownership of the franchise covering the southwestern portion of Dayton city.

CABLE SUBSCRIPTION FEES

- *How will subscription fees be determined? Will uniform fees be charged to all residential subscribers? How should fees vary between residential subscribers and institutional users?*

In virtually all present cable systems, most of them rural, subscribers pay a standard monthly fee. As cable television enters highly urbanized areas, however, the fixed-fee structure may prove unnecessarily discriminatory. The greater the residential density, the lower the costs of wiring an area for cable television, which means that, all other things being equal, it will cost less to wire the central city than to wire suburbs. (The cost difference could be offset, however, by a higher penetration level in the suburbs, or by having to wire underground in the central city.) Consequently, a fixed-fee structure for a metropolitan-wide system may discriminate against the central city resident. In effect, he could be subsidizing his suburban counterpart in a region-wide cable system.[43] The problem is potentially all the more serious since suburbanites generally have higher incomes than central-city residents.

If fixed fees are discarded, the question is how to determine a differential fee structure. One solution is to have the fees for the central city and the suburbs reflect the actual cost of wiring these areas. The cost-differential between wiring single-family homes and apartments can also be reflected by differential rates. Citizen participation, particularly involving representatives of both inner-city and suburban areas, could be helpful in determining other criteria and ultimately the most equitable fee structure. Naturally, this issue is highly complex. It finds direct parallels in the fee structure for the telephone system and other utilities. Whether cable television should follow patterns set by those other industries or whether new patterns should be established must presently be decided independently by each municipality.

Then there is the equally important question of the variation between residential and institutional users. Present plans for new urban cable systems call for a whole array of public service uses in addition to new entertainment.[44] Some, such as home education and other programs providing public interest information to the resident, will be directed at the individual subscriber. Others, however, such as improved classroom education and medical diagnosis via television, will be directed only at the internal communications of a particular municipal agency. In some cases, these internal communications may require a large amount of channel space.

According to the 1972 FCC rules for cable television, any significant institutional involvement in the cable system will be on a leased channel basis. This raises the question of what fees institutions should pay and how the fees should be determined. It would seem inequitable, for instance, if the cable system were supported primarily by residential subscription fees even though the heaviest users were public or private institutions. On the other hand, institutions operating public services could conceivably claim that they should pay lower fees, since the beneficiaries

---

[43] This conclusion was reached by the Rand staff in the course of Rand's study of Dayton, Ohio, and explicitly noted in one of the early working papers. See Robert K. Yin, "Television and the Dayton Resident," preliminary paper, October 1971.

[44] For a review of these potential services, see the companion reports in this series by Polly Carpenter, "A Guide for Education Planners," and Robert K. Yin, "Applications for Municipal Services," chapters 4 and 3, respectively, of Carpenter et al., op. cit.

of any improved agency communications will still ultimately be the residents, who are the clients for those public services. Again, similar arguments on both sides have been made in relation to the pricing of telephone service. And again, citizen participation in the planning stage could identify additional pros and cons of this debate and could help to develop the most acceptable solution.

CABLE SERVICES

- *What types of services and programs are desired? How many channels should be dedicated for special use? Should the television set itself be serviced by the cable operator?*

Advanced cable television systems allow for a wide variety of television uses, well beyond over-the-air television. Cable television can bring two-way interactions between residents and a central studio, special channels set aside for municipal and community programming, and dedicated channels in which only users with special equipment are able to receive the programs. A dedicated channel, for instance, might be created for continuing medical education for doctors, where the programs might be too specialized or possibly even offensive for the ordinary viewer. Still another special arrangement might be the creation of one or more pay-TV channels. These channels could show especially attractive programs, be available only to viewers who would pay extra money to see them, and add to the cable operator's revenue.

The FCC has stipulated certain minimum requirements for channel capacity and usage, requiring, for instance, that channels be set aside for education, local government, and public access. However, such requirements do allow for some choices to be made by municipalities, and citizen participation would serve as an appropriate forum for further discussion, even to the point of recommending changes in the FCC regulations. Adequate consideration in the planning stages might suggest additional requirements that should be written into the franchise agreement. Citizens could help decide, for instance, how the total channel capacity should be allocated for different uses, and how many channels ought to be leased for public services or for pay-TV. Citizens could also discuss the prospects for interactive services, such as facsimile mail, shopping via television, and take-at-home school exams, and suggest ways to help develop these services in the community.

MONITORING CABLE OPERATIONS

- *How should cable operations be monitored to insure that franchise provisions are enforced?*

Citizens can suggest what mechanisms, if any, should be used to monitor the cable system in operation, both for program quality and relevance, and for any conditions stipulated by the franchise. For instance, certain guidelines may be needed on how to deal with citizen complaints over service deficiencies. To date, however, very few municipalities have developed a satisfactory means for monitoring cable operations to see that these provisions are fully satisfied. New York City, for instance, has created an Office of Telecommunications that is charged with monitoring the two cable franchises operating in Manhattan. Citizens may want to suggest a similar focus of responsibilities in their communities.

IV. PLANNING FOR CABLE TELEVISION
WITH CITIZEN PARTICIPATION

Section III presented several topics for citizens to consider, discuss, and decide upon in relation to cable-TV planning. This section deals more with the mechanics of participation: when it can begin and what forms it can take. Much of our discussion will draw on recent research experiences in planning for advanced cable systems in Dayton, Ohio.[45] Before beginning, however, an earlier caveat is worth repeating: general guidelines for citizen participation can only be suggestive; the very spirit of citizen participation dictates that the ultimate form and content of citizen participation can be decided only by the people and elected officials in each locale, and only according to their interpretation of what the local situation requires. In many cases, state or local regulations may already call for specific forms of citizen participation, and local citizens should become aware of such regulations.

WHEN PARTICIPATION MIGHT BEGIN

At first glance, the formal starting point for citizen participation would seem to be merely one among many similar steps in the participation process. The starting point is probably the most critical step, however, for the initial encounters and interactions will set the tone for the remainder of the process. Of major importance at the outset are questions of style and form, rather than content. For instance,

- What local personnel or institutions should take the initiative for planning cable television?
- What has been their past role in community affairs?
- What is the initial style of communication (e.g., pluralistic or authoritarian) about cable television within the community?
- What commitments, if any, have already been made?

And, at the beginning, one of the first overt signs to local citizens will be the establishment of citizen participation, whether through the call for a formal citizen group, or for informal participation. The timing of this step, above all, will set an

[45] For the full report of the Dayton study, see Johnson et al., op. cit.

important precedent and provide the clues for the likely outcome of the cable planning process.

Ideally, every community that wants to award a cable television franchise will engage in a preparatory phase before holding public hearings, making binding commitments, or considering specific franchise terms. This preparatory phase is the most appropriate juncture for participation to begin.[46] It should allow full exploration of the social, economic, and technical aspects of the cable system, beginning with the most important question of all: Should a cable system be built in the municipality at this time? In considering a cable system for the Dayton metropolitan area, this preparatory phase took more than nine months, during which a study team examined a wide range of issues concerning potential cable systems for Dayton. During the entire time, a self-policing moratorium was in effect, under which none of the municipalities in the Dayton metropolitan area granted cable franchises for their jurisdictions, although each was empowered to do so. Other communities may not have the time or resources to support a preparatory phase in the Dayton style. However, new national organizations, such as the Cable Television Information Center in Washington, D.C., can facilitate the preparatory phase by providing prompt and relevant advice and guidance to specific municipalities. The Cable Bureau of the FCC may also provide a field team to answer questions on franchising.

There is another way, however, of defining the point at which participation should begin. The definition is related to the participatory process rather than the cable planning process, and stems directly from the recent urban experiences with federal programs. According to this definition, participation begins not by calling citizens together for deliberations about cable television, but at a distinctly prior step: Citizens must be involved in deciding *how* they are to participate in the planning process.[47] Thus, they begin with discussions concerning the format and process of citizen participation itself. Furthermore, this discussion can start even before any formal study group is engaged to carry out the research appropriate to the preparatory phase. Since an advanced urban cable system is supposed to be responsive first and foremost to the citizens, a good indication of this intent is to start citizen participation before negotiations occur with parties outside the community, whether they are research, foundation, or university groups, cable television operators, or potential financiers. Citizen participation would then mean that citizens could help to design the entire preparatory phase as well as the cable system itself.

WHAT FORMS CITIZEN PARTICIPATION MIGHT TAKE

Local experience with federal programs has led to many forms of participation. While the forms are varied, it is possible to consider them all as lying on a single conceptual dimension, one having to do with power relationships. Certain types of participation, such as open meetings held merely for one-way informational pur-

[46] See footnote 2, and Chapter 2 of this book.

[47] Edgar S. Cahn and Jean Camper Cahn, "Citizen Participation," in Hans B. C. Spiegel (ed.), *Citizen Participation in Urban Development*, NTL Institute for Applied Behavioral Science, Washington, D.C., 1968, pp. 211-224. The authors also point out the logical problem of infinite regress in this issue.

poses, allow for little or no power to be exercised by citizens; other types, such as citizen groups created by specific delegations of authority, can encourage considerable citizen control.[48] The Dayton experience suggests that the type of participation will be a function of the level of community development, signified by the community's ability to organize and sustain interest and leadership on a given topic. If there is a high degree of development, community groups will be easy to organize (or will already exist), and they will automatically seek a certain amount of control over the situation, no matter what type of participation is planned at the outset. Weak community development will mean difficulties in organizing and fewer citizen initiatives, again regardless of the initial type of participation. In short, the main determinant of the form of participation is the citizenry's ability and readiness to organize. This ability will vary from one community to another. For example, an important side-effect of the Anti-Poverty and Model Cities programs has been the significant improvement in community organization in low-income inner-city neighborhoods. As one observer has noted concerning the Anti-Poverty program,

> Organizations now exist that might never have come into existence without the impetus of OEO. These are new organizations—not funded by OEO but originally organized in response to issues created by OEO programs.

> The Zeitgeist—the atmosphere and mood of the ghetto—has changed, and is not likely to regress to its former apathy and resignation.[49]

At the same time, communities that have had no active part in these federal programs are likely to be more difficult to organize.

As a minimum requirement, regardless of which form citizen participation takes, a franchising authority should provide a procedural framework that guarantees citizens meaningful notice and opportunity to be heard at each step of the franchising process, from the initial determination of franchise areas to the final award of the franchise. Such a procedural framework is discussed in more detail in Chapter 5 of this book.[50]

Other forms of citizen participation may be of at least four types: (1) surveys, (2) conferences, (3) ad hoc committees, and (4) delegated groups. The first two types require little community organization, while the last two assume some ability to sustain organized participation.

Surveys

Surveys are one of the few methods whereby the reaction of the everyday citizen can be tested. In most other forms of citizen participation, the representativeness of the community leaders can always be questioned. A well-designed survey, on the other hand, will seek interviews from a sample of residents that represents every segment of the community. Because of this, and because surveys are relatively easy and inexpensive to carry out in most communities, more detailed discussion will be devoted to this form of participation than the others.

---

[48] For more examples and the potential underlying typology, see Arnstein, op. cit.
[49] Lazar, op. cit., p. 107.
[50] See pages 74-92 of this book.

What Surveys Can Accomplish. The first and possibly most important function of a survey is that it solicits the opinion of the ordinary citizen about cable television, and suggests that he may have some influence on the final design of the cable system. The act of carrying out a survey can also furnish a broader base of local support for cable systems. In most cases, municipalities have called for feasibility and planning studies by technical groups located *outside* of the municipality; the survey is thus one of the few means of involving local citizens. Therefore, for their participatory value alone, surveys may be considered a useful part of cable system planning.

Surveys can also provide helpful information for planning the cable system.[51] The first category of such information involves the amount of usage of television. A survey can determine, on a small-area or metropolitan basis, answers to the following sorts of questions:

- How much television is currently viewed?
- Who watches television?
- What programs are of the greatest interest?
- How much subscriber interest is likely to exist for cable television?
- How much subscriber interest is likely to exist for various types of new programs, as well as for various types of new interactive and public interest services?

A second category of information has to do with the communication patterns among residents. Certain districts within a city may have more frequent communications with each other, certain institutions may need to communicate more frequently with residents of specific districts, and certain segments of the city population may desire greater communication with residents in other cities. A survey can identify these patterns, and, if carried out properly, can actually provide some guidelines for the placement and interconnections of the cable hubs and trunk lines. It can also explore the preferred patterns of communication, not merely the existing patterns. For instance, a recent survey has examined the communication patterns among different agency officials in a metropolitan area.[52] Analysis of these patterns yielded clues to the pattern of interconnections and the relevant type of studio equipment needed to facilitate these agency communications.

Procedures for Conducting Surveys. The general procedures for conducting urban surveys have been outlined in an excellent booklet by The Urban Institute.[53] The booklet reviews the fundamentals of questionnaire design and sampling procedure, and estimates the likely costs. It also weighs the advantages of different survey methods (mail, telephone, and personal interviews). The booklet is recommended as a reference and guidebook for any municipality planning a survey or

[51] For a full discussion of the pertinent issues that surveys can cover, see Robert K. Yin, "Cable on the Public's Mind," *Yale Review of Law and Social Action,* Spring 1972. See also the survey carried out by the MITRE Corporation in its study of Washington, D.C. (William F. Mason et al., *Urban Cable Systems,* the MITRE Corporation, Washington, D.C., May 1972.)

[52] See Daniel J. Alesch, *Intergovernmental Communication in the New York-New Jersey-Connecticut Metropolitan Region,* The Rand Corporation, R-977-MRC, May 1972.

[53] Carol Weiss and Harry F. Hatry, *An Introduction to Sample Surveys for Government Managers,* The Urban Institute, Washington, D.C., 1971.

needing more information about surveys. Two additional procedural points should be made with regard to surveys for cable television.

First, the results of a survey can be biased by the types of questions asked (and not asked), and by their wording. Questions should also be relevant to the locality, meaning that the survey team must have some knowledge about local issues and interests. For these reasons, the choice of the survey team and the questionnaire design are extremely important.

Most municipalities do not conduct their own surveys. Residential surveys are usually contracted out to a university-based opinion research team. The procedures for selecting the survey team should be made as openly known as possible, and, if feasible, the survey should be conducted by a locally based group. Such a group is more likely to design questions that are most meaningful for both cable issues and local issues. Where possible, citizen groups may be consulted on the choice of the survey team and the design of the questions to be asked.

Secondly, a survey for a metropolitan area of up to one million persons can be carried out with a sample of about 700 residents, so long as the sample is selected carefully and according to appropriate statistical procedures. This means that a survey involves only a moderate financial investment of about $10,000 for personal interviews in most medium-sized cities. (The sample may have to be larger, however, if very fine stratifications of respondents, e.g., according to detailed geographic subsection, age, race, or residential breakdowns, are desired.) Furthermore, since the same survey offers the opportunity to ask questions about municipal issues other than cable, such as the use of or attitudes toward various services, the survey can serve many purposes simultaneously. Several municipal agencies might therefore want to share the cost of the survey.

Problems Posed by Surveys.   Although the benefits of surveys seem promising, and although the costs of surveys are generally low, municipal officials and other participants in the franchising process should approach a survey carefully. This is because carrying out a survey entails certain risks that could threaten the development of the cable system. First, a survey will be most useful if conducted early, before the cable franchise has been negotiated. This is to allow for the development of a franchise that will incorporate any relevant conditions identified in the survey. Such an early survey, however, may mean the premature disclosure of the development of the cable system. In other words, residents may need special education about the potential of cable television before they can react to it; without such education a survey may arouse unnecessary fears.

Second, a survey could inflate citizens' expectations about the amount of influence they will have over cable-system design. If that occurs, and the results of the survey appear to be ignored, they may react with unnecessary hostility to any cable system. It is also possible that the survey will yield unexpectedly negative results: citizens may not favor the construction of a new cable system, especially in comparison with other alternatives. Municipal officials must decide beforehand if they will abide by such a negative outcome, and, if not, how they will deal with the situation. One obvious recourse is simply to drop the idea of having a cable system.

Finally, as a form of citizen participation, surveys are an extremely passive device. The citizenry at large has no control over the design of the survey questions, over how the survey is ultimately to be used, or over any follow-up measure. In short, the survey elicits information from the everyday citizen and gives him a voice in the cable planning process, but obviously does not give him any power or control.

Conferences

Conferences serve the purpose of briefing the public and informing them of the major issues concerning cable television, but do not require citizens to raise any questions themselves. Conferences can communicate research findings or cable plans to wide audiences, and allow for substantive discussions and interactions.[54] They can be held frequently and before many different audiences.

At conferences, however, citizens are distinctly subordinate to those who are actually directing the cable planning process. Most conference participants have not been exposed to enough information beforehand, and so cannot really become intelligent and constructive discussants. Furthermore, a conference does not endow them with any control over the cable planning process itself. The conference, in other words, acts as a complement to surveys: whereas surveys emphasize the flow of information from the citizenry to cable planners, conferences emphasize the opposite flow. But in neither case is there any citizen control or power over cable television planning or operation.

Ad Hoc Committees

Ad hoc committees are composed of citizens representing various interest groups, whether such interests are defined functionally (public health, education, etc.), or geographically. Given the proper resources, there may be many committees, and each can meet several times during the preparatory phase. The longer their exposure to cable television issues, the greater their potential for directing the course of the planning. In the Dayton research, ten committees were formed, representing the following interest groups: public health, education, technical and engineering, government, religion, community development, arts and culture, law and finance, medicine, and medical education. These committees met individually several times during the nine-month preparatory phase. Each meeting began with a briefing that summarized the study to date; then the members raised questions and made recommendations that were later incorporated into the feasibility study. The study thus benefited from the expertise of a wide variety of people, some of whom spoke as potential users of the cable system and others of whom had experience in communications and mass media, but all of whom could speak with the experience of having dealt with local affairs.

In and of itself, the formation of these committees does not imply a powerful type of citizen participation. Participation may still be limited to advice-giving. However, the ad hoc committees provide an opportunity for citizens to exert greater influence over the planning process if they so desire. First, by meeting several times, the committees can make demands and pursue them until they are satisfied. Second, with adequate financial support, the committees can develop their own sources of information, such as conducting their own surveys about cable television. Third, the committees can act as advocates for their interest groups. In other words, in spite of their ad hoc nature, the committees may be able to establish a certain degree of representative legitimacy. Fourth, the committees can continue beyond the plan-

[54] In Dayton, municipal officials and community groups were invited to a "Policy Makers Conference on Urban Cable Communications," held on January 25-26, 1972.

ning process and support particular issues during the development and granting of the cable franchise. In each case, citizens must decide the extent to which they will take the initiative and use these opportunities to make the ad hoc committee structure more influential.

Delegated Groups

Of the four forms of participation, delegated groups have the greatest potential for citizen control over the design and development of the cable system. Delegated groups are citizen groups that have been given specific authority by municipal officials to make independent assessments during the planning process, to speak on certain subjects in franchise negotiations, and to monitor the cable system in its construction and operation. In the planning process, the activities of delegated groups can parallel those of any outside study team. They can cooperate with the study team, conduct their own research, or even replace the study team. To be effective, however, a delegated group must have an assured part in the franchise negotiations; otherwise, its influence on the planning process may be minimal. Having such a continued role might also induce the delegated group to act more responsibly, and prevent situations in which a group strongly supports certain issues on one day, and then disappears on the next. One good sign of the relative power of delegated groups is the extent to which they dictate the planning timetable. The more influence they have, the more they will be able to set rather than have to meet deadlines.

In the Dayton experience, the delegated form of participation was not explored. It could have been, had the ad hoc committees or existing community organizations been given actual authority to carry out part of the planning process. As a rule, most cities may not want to use the delegated group structure, as it requires considerable commitment from both municipal officials and the local citizens. The officials must delegate authority and also provide resources (both money and staff time) to the citizen groups so that they can carry out their work properly; the citizens, on the other hand, must be willing to pursue the planning process to its conclusion, and not fail to carry out their full responsibilities within reasonable time constraints. For many people, this can mean sacrificing a large amount of time for which other priorities compete, such as their participation in other community affairs. Here it is important to note that cable television, in spite of its great potentiality for reshaping human communication patterns, must still compete for attention with other important social events and programs. In other words, any anticipated continuity of support for cable planning must be viewed within the context of the overall activity in the community; the continuity may easily be broken by new developments in federal programs, community crises, and any number of other events that may appear more immediately important than cable television.

V. CONCLUSION AND RECOMMENDATIONS

Three groups have dominated the development of cable television in the cities: cable operators who have accumulated experience by building cable systems in rural areas; television networks and owners of over-the-air television stations who have acted to prevent the expansion of cable television into the top urban markets; and federal policymakers and technical advisers who have guided FCC policy. There is a logical fourth participant, however—the urban citizen himself, who is alleged to be the prime beneficiary of cable television.

The call for citizen participation in the development of cable television, and especially in its planning, comes from the recognition that urban cable development will occur within the broader context of recent events in the American city. These events have included consistently greater citizen involvement and control over activities affecting the city, primarily over such federal programs as urban renewal, Anti-Poverty, and Model Cities. Cable television differs considerably from these federal programs in being a private, not a public enterprise. Nevertheless, it will arouse similar levels of citizen interest because the new television services can play such an intimate and important role in human communication systems, especially in the development of the inner city.

This report has described some of the historical roots of citizen participation in the cities as well as some of the likely reactions of citizens to cable television. Citizens are likely to view the development of cable television with suspicion, they may expect recognition as experts on the planning and design for serving community needs, and they may even obstruct development of any cable system if community ownership and control are not considered as alternatives. The report has also discussed several of the major community-related issues in cable TV planning: ownership, cable system geography, subscription fee structure, cable services, and monitoring of cable operations. Finally, the report has suggested the most appropriate timing for starting the participatory process and described several forms of participation and their advantages and disadvantages.

The line of argument leads to the following recommendations:

• Municipalities should allow a reasonable period of time (perhaps a year for a medium-sized city) between the announced intention to develop cable television and the first public hearings regarding the actual franchise terms.

• The period should not be construed as a device for excessive delay. Rather,

255

municipalities should use the period to encourage and support the formation of citizen groups to plan, review, and advocate cable-related issues.

• This type of citizen involvement should begin as early as possible in the planning process, before major options have been foreclosed.

• Citizen involvement or participation can take many forms, most notably conferences, surveys, ad hoc committees, and delegated groups. The form, timing, and style of citizen participation should be determined as much as possible by the citizens themselves.

• Citizens should feel free to consider any cable-related topic or issue, including: geographic coverage; subscription fee structure; cable system design; cable ownership, operation, and monitoring; access to public channels; education and municipal channels; and the types of services to be provided.

In summary, the successful development of cable television in contemporary American cities depends on a well-coordinated effort in planning for cable television in each municipality. Such an effort should also be an open affair, in which all groups and citizens can participate if they choose. To make the planning stage a pluralistic process obviously entails considerable risks. Citizen participation can lead to delays, conflicts, or even opposition to the development of any cable system. These risks need to be taken, however, if the cable system purports to serve the public interest. This is one of the few ways in which cities ultimately derive the greatest benefits from the new cable systems, whether these involve community ownership and control, improved public services, new and attractive types of entertainment, or expanded opportunities for communication among people and institutions.

Chapter 5

Citizen Participation After the Franchise

Monroe E. Price and Michael Botein

I. INTRODUCTION

As cable television systems grow across the country, most opportunities for creative contribution by local groups will be at the municipal level, when franchises are shaped and awarded. But it would be a great mistake to overlook the power of the Federal Communications Commission to assure that cable system operations will serve the public interest. In 1972, the FCC established rules for the growth of cable television, setting very specific federal standards.[1] These standards encourage experiments in the use of cable television for education, for local government uses, and for public access to the medium. The federal standards also assure minimum technical standards and require public hearings during the franchise award process.

Aside from the rules themselves, the FCC's main instrument of regulatory power is the *certificate of compliance*. Before a cable system can carry broadcast television signals, it must apply for and obtain such a certificate from the FCC. The process leading to the award of a certificate of compliance is still new, still being shaped. It is an important step in nurturing a responsive cable television system in the community, since it provides the principal opportunity for federal review of local decisions, and since it constitutes yet another forum in which interested parties can air their views.

Already, broadcasters have begun to use the certification process in attempts to influence FCC policy and cable practices. But community groups and local governments have largely neglected the same opportunity. As of this date, the FCC has approved roughly 1200 certificate applications. According to an FCC staff member, in not one case has a community filed an opposition. This report explains the certification process and its relationship to the franchise. The intent is to help groups understand how they should examine an application for a certificate of compliance and what actions they should take before the FCC.

In addition to the certification process, local groups may apply to the FCC for special relief at any time during the franchise period, and may comment on applications for waivers and reliefs sought by others. The procedures for doing so are discussed in Sec. VII.

To understand the complexity of the certification process, some understanding of the FCC rules is important. Section II furnishes the necessary background.

---

[1] The FCC rules, embodied in the *Cable Television Report and Order*, FCC 72-108, appear in the *Federal Register*, 37 Fed. Reg. 3252 (1972); the *Reconsideration* appears in 37 Fed. Reg. 13848 (1972). Both documents are reproduced in a companion report in this series, Steven R. Rivkin, *Cable Television: A Guide to Federal Regulations*, Crane, Russak & Co., New York, 1974.

II. BACKGROUND OF THE FCC RULES

Citizen participation is essential not only in drafting a franchise and selecting a franchisee, but at every decisional stage in the franchising process. Adoption of a franchise does not mark the end of participation by local governments or citizen groups. A cable operator needs the FCC's permission to carry broadcast television signals, and his subsequent performance may require explicit FCC approval or may incur its disapproval. Both community groups and local governments must actively intervene in these processes.

The FCC adopted its new cable television rules by way of its February 2, 1972 *Cable Television Report and Order.* The rules are the product of a prolonged running battle among cable, broadcast, and copyright interests. As a result, they focus on these groups' main concern—cable's use of broadcast television signals. Local governments and citizen groups must also be familar with them, because some aspects of the rules will directly affect a cable system's service to its community. More important, they also restrict a franchising authority's power to regulate a cable system. Finally, the new rules create several difficult procedures for modifying their requirements. The following brief review may be helpful to local governments and citizen groups, but serious matters call for careful analysis of the rules or consultation with an expert communications lawyer. (See Rivkin's *Cable Television: A Guide to Federal Regulations* in this series for a detailed analysis and commentary on the FCC regulations.)

FCC RULES ON CABLE USE OF BROADCAST TELEVISION SIGNALS

Like the FCC's previous and proposed rules, the new regulations concerning cable use of broadcast television signals concentrate on two main provisions: first, that a cable system *must* carry some signals; second, and more important, that it *may* carry certain other signals. (In this context, broadcast signals are those picked off the air by antennas and retransmitted over cable; nonbroadcast signals are transmitted entirely by cable.)

Sections 76.61 and 76.63(a) require a cable system in the 100 largest television markets to carry the signal of (1) any television station within 35 miles of the cable system's community, (2) any educational station that places a grade B contour—that is, a moderately receivable signal—over the cable system's community, (3) any commercial "translator," i.e., relay station with 100 or more watts of power, (4) any station licensed to the cable system's market, and (5) any "significantly viewed" station. Under Sec. 76.59(a), the standards for markets below the top 100 are essentially the same, except that a cable system there must carry any commercial as well as noncommercial station that places a grade B contour over its community. Finally, Sec. 76.57(a) requires a cable system outside of all markets to carry a station that meets any of these standards.

The regulations limit the *number* of distant signals a cable system may import. Thus, cable systems may import enough distant signals to offer "minimum service" —i.e., three network and three independent signals in the top 50 markets, and three network signals and one independent signal in the smaller markets. (See Secs. 76.61(a), 76.61(b), and 76.51(b).) In addition, Secs. 76.61(c) and 76.63(a) allow major market cable systems to import two additional "wild card" independent signals, unless these already have been used to provide "minimum service."

Second, a cable system must choose the geographic source of its distant signals in accordance with "leapfrogging" rules. If a system wants to carry signals of stations in the first 25 major markets, it must choose one or both of the nearest two such stations if they are available. And if a system can play a "wild card" for a third independent distant signal, under Sec. 76.61(b)(2), it must look first to any independent Ultra High Frequency (UHF) station within 200 miles or, if none is available, any independent VHF station within 200 miles or an independent UHF station anywhere.

Third, and most important, the regulations give local stations "exclusivity rights" that prevent a cable system from carrying a particular program. Thus, even though authorized to import a distant signal, a cable system may not carry programs that a local broadcaster has the exclusive right to show. Section 76.151(a) provides that a cable system in the top 50 markets may not show a syndicated (nonnetwork) program for "one year from the date that the program is first licensed or sold" to any station—even if not bought by a local station. In addition, under Sec. 76.151(b), a cable system may not import a program that any local station is showing if the contract between the broadcaster and the program supplier provides for exclusivity. In the 50 next largest markets, the exclusivity provisions are somewhat less stringent. Section 76.151(b) does not impose the one-year presale ban, although it allows very extensive contractual exclusivity rights. A station may contract with a program owner to bar syndicated reruns for up to a year and to prohibit new syndicated series programs, nonseries programs, and feature films for up to two years.

FCC RULES ON USE OF NONBROADCAST CABLE CHANNEL CAPACITY

The regulations contain a number of provisions requiring substantial channel capacity for uses other than carrying broadcast signals. First, and perhaps most important, Sec. 76.251(a)(1), (2) requires that all major market cable systems have as many channels for nonbroadcast as for broadcast use, with a *minimum* capacity of 20 channels. As a result, all major market systems must have at least 20 channels, and some systems in the larger markets—such as New York, Los Angeles, and Chicago—may need to provide 30 or more channels. Moreover, Sec. 76.251(a)(3) requires all major market cable systems to build in at least some capability for two-way communication between subscribers and the system. The regulation is vague, however; for the time being, it will require operators to do no more than lay cable that can accommodate two-way communication at some undefined point in the future.

Second, Sec. 76.201 retains the Commission's previous requirement that cable systems with more than 3500 subscribers originate programming "to a significant extent as a local outlet"—a regulation the Supreme Court recently upheld in the landmark *Midwest Video* case (the decision is reprinted in full in Rivkin, op. cit.). As will be discussed later, the meanings of "local outlet" and "significant extent" are less than clear; however, the regulation appears to require a cable operator to operate somewhat like a local television broadcast station.

Third, the regulations specifically provide for both free and leased use of cable systems' channel capacity. Section 76.251(a)(4)-(6) requires major market cable systems to provide one "public access," one "education access," and one "local government access" channel for free. Section 76.251(a)(11) provides that major market cable systems must offer to lease all unused channel capacity. Finally, Sec. 76.251(a)(8) imposes the requirement—previously known as "N+1"—that a cable system add a new channel whenever all nonbroadcast channels "are in use during 80% of the weekdays (Monday-Friday) for 80% of the time during any consecutive three-hour period for six consecutive weeks."

Unfortunately, the access rules are filled with ambiguities and pitfalls. Because the FCC has not defined "public," "education," or "government" access, it is difficult to say who is entitled to use a particular channel. In addition, the FCC has left control of the access channels largely to the unfettered discretion of the cable operator. Thus Sec. 76.251(a)(11) requires simply that *systems* "shall establish rules requiring first-come nondiscriminatory access." And to compound the confusion, the FCC specifically prohibits local or state governments from stepping in to fill this regulatory void. Section 76.251(a)(11) explicitly states that a local government may not impose "any other rules concerning the number or manner of operation of access channels." Moreover, the FCC has offered virtually no help in drafting access rules; its standard of "first-come nondiscriminatory" access is almost meaningless. Finally, the access rules impose contradictory duties on cable systems. Section 76.251(a)(9) provides that cable systems "shall exercise no control over program content" on free or leased channels. But Sec. 76.251(a)(11) then requires cable systems to exclude some types of programming, such as obscenity, lotteries, and the like.

FCC RULES ON FEDERAL, STATE, AND LOCAL REGULATORY RELATIONSHIPS

The FCC's new regulations do not define regulatory roles for federal, state, and local entities, but rather restrict the powers of local governments. They lay down a set of fairly stringent—albeit also fairly vague—standards for local governments in granting and administering franchises.

The rules do not, however, directly impose any requirements on local governments. Instead, they require a cable system to state that its franchising authority has followed the FCC's standards as a prerequisite to obtaining permission to carry broadcast television signals—the FCC "certificate of compliance," discussed in the next subsection. This approach presumably stems from the FCC's very legitimate doubts about its power to impose obligations directly on local governments. But since few franchising authorities will wish to prevent their cable systems from becoming operational, most governments will hew to the Commission's line.

First, Sec. 76.31(a)(1) requires a local franchising authority to consider a cable operator's "legal, character, financial, technical, and other qualifications" by way of a "full public proceeding affording due process." Similarly, under Sec. 76.31(a)(4) a franchising authority must "specify or approve" the "initial rates" charged *subscribers,* as opposed to users; moreover, a cable system may not change its *subscriber* rates without authorization from the franchising authority "after an appropriate public proceeding affording due process." As will be discussed later, the meaning of a "full" or "appropriate public proceeding affording due process" is obviously less than clear. In addition, Sec. 76.31(a)(5) requires the franchise to "specify procedures for the investigation and resolution of all [subscriber] complaints."

The rules also impose fairly lenient construction deadlines upon cable operators. Section 76.31(a)(2) requires a cable operator to "accomplish significant construction within one year" of receiving his certificate of compliance, and provides that he must wire "a substantial percentage of its franchise area each year" in an "equitable and reasonable" fashion. Under Sec. 76.31(a)(3), an initial franchise may not be longer than 15 years and a renewal franchise must be of "reasonable duration."

Finally, and in many cases most important, Sec. 76.31(b) limits a local government's franchise fee to 3 percent of a system's gross annual receipts from *subscriber* services. A local franchising authority may charge a higher fee only with the FCC's specific approval, and the FCC will approve a higher fee only upon a showing that it "will not interfere with the effectuation of the federal regulatory goals... [and] that it is appropriate in light of the planned local regulatory program." As will be discussed later, both the form and substance of the necessary special showing remain unclear.

The FCC has passed on few certificate applications from new systems to date. Its actions with respect to older grandfathered systems, however, suggest that it will not enforce its franchise standards very stringently.

III. THE CERTIFICATE OF COMPLIANCE

Section 76.11 of the FCC's new rules provides that "No cable television system shall commence operations or add a television broadcast signal to existing operations unless it receives a certificate of compliance from the Commission."

To receive a certificate of compliance the franchisee must submit to the FCC a fairly elaborate document that contains, among other things, statements explaining:

1. How the system proposes to fulfill the goals of federal regulations establishing a channel for education uses and a channel for local government uses.
2. How the system's plans are consistent with federal regulations requiring the establishment of a special channel for public access—that is, the right of individuals and groups to reserve time for the distribution of their own television programming.
3. The system's plans for meeting federal requirements that systems in large markets[1] locally originate to a significant extent and that systems in small markets provide certain programming facilities.
4. How the franchise fulfills the federal expectation that cable systems will be built on time and that cable will be equitably spread throughout the franchise area.
5. How initial subscriber rates were set, and the procedure to be followed when subscriber rates are to be amended.
6. How the franchise process met standards requiring a "full public proceeding affording due process."

Thus, a cable system may not broadcast additional television signals unless the FCC has approved its franchise by way of a certificate of compliance. Of course, a cable system that did not carry the popular off-the-air television stations could operate without a certificate. But realistically, a cable system needs a certificate to be economically viable. Even though cable's future may depend more on new services rather than broadcast signals, no present system can afford to operate without broadcast signals and an FCC certificate.

There is no set form for an application for a certificate. Most applications filed since the rules went into effect on March 31, 1972 have been sketchy; nor is it yet

---

[1] The "top 100" or "major" television markets.

clear how thorough the FCC, through its Cable Television Bureau, can be in reviewing the applications. Thousands of applications will be filed, and the staff of the Bureau is small. The only penetrating analysis will come from community groups that monitor applications. And only as groups file before the FCC, listing inadequacies in application forms, will the FCC act directly and frequently to insure the integrity of the certification process.

The FCC developed this process as an alternative to direct federal licensing of cable television systems or complete reliance on local franchising authorities. Under the rules, the federal government establishes standards for performance—standards the local community is obliged to follow in awarding the franchise. The certification process is designed to assure that the franchised cable system actually meets federal standards.

The application process is also an opportunity to justify deviations from federal standards. In this report, specific attention will be paid to reviewing applications that seek to justify special terms, such as franchise fees higher than 3 percent, more than three reserved free channels for public use, and special controls over originated programming.

The application for a certificate of compliance should be ample, thorough, and descriptive. It is likely that even a precise and complete cable franchise will merely set norms without describing in detail how the norms are to be reached. The application for a certificate can be more generous. One example is the matter of access channels. The FCC rules require the cable operator to set aside one channel for education uses. The franchise will normally affirm this administrative prerequisite. The application for a certificate can go beyond reiterating the requirement, and tell in greater detail how the system hopes to implement the federal goal of instructional use of part of the cable capacity.

IV. GETTING ACCESS TO THE APPLICATION FOR A CERTIFICATE OF COMPLIANCE

Community groups that wish to comment on an application for a certificate have a threshold problem: obtaining the document in time to analyze it thoroughly and make useful comments. Though the rules specifically allow interested persons and groups to file objections to certification applications, community groups must move quickly to be heard. Section 76.13 requires a cable system to serve a copy of its certification application only upon the franchising authority, local television stations, the superintendent of schools, and a few others. Under Sec. 76.15, the Commission will "give public notice of the filing of applications"; but community groups that do not scrutinize the FCC's releases or the *Federal Register* will find themselves in an informational vacuum. The rules also require that the cable operator make a copy of the application publicly available if the franchising authority does not.

The time allotted for community comment on the certificate application is extremely short. Unless an extension is obtained, a group must file comments and objections within 30 days of public notice by the FCC. This means that interested groups and persons must be vigilant during the postfranchise period to make sure they receive a copy of the application.

Some definite steps can be taken to ensure that community groups have an opportunity to express their views on a certificate application. The franchise itself could require the cable operator to give more widespread notice, perhaps through local newspapers or cablecast announcements over the system.[1] Another solution would require the cable operator to submit his application for FCC certification to the franchising authority and publish it before the franchise is granted. Such early drafting would allow a reasonable time for discussion and revision—well in advance of the 30 days allowed by the FCC. In many cases it will be wise for community groups, the franchising authority, and the franchisee to draft the application together and thus reduce the likelihood of friction during the certification process.

[1] For a more detailed discussion of these and other methods, see Chapter 2 of this book.

V. ANALYZING AN APPLICATION

Now that the interested group has an opportunity to review the application, how should the document be analyzed? How thorough should an application be? How should a community group draft its comments and objections for submission to the FCC?

Community groups should ensure the accuracy and completeness of certification applications for a number of reasons. Section 76.13 requires a cable operator to state how he will comply with many of the FCC's regulations. Though statements in an application may not be as binding as promises in a broadcast license application, they can be a good yardstick for measuring the operator's later performance. What he promises will be important at renewal time and perhaps can be the basis for subsequent petitions for special relief.

His promises should be clearly stated. Community groups should ensure that they are spelled out fully and not hedged with cautious lawyers' language. For example, the cable operator's statements about planned program origination call for careful review. If his franchise application promised high-quality children's programming, his certification application should reiterate the promise and explain how he will fulfill it. The rules require a cable system to disclose a great deal about its ownership and financial structure. Community groups should review this information. They should check its accuracy against data the cable operator filed as part of his franchise application. They should examine it to determine his financial ability to carry out his promises. They should make sure it is complete and clear. Ownership disclosures may suggest possible conflicts of interest, concentration of control, or lack of minority participation. The certificate application provides an opportunity to explain how the local system intends to meet or surpass federal standards. If the cable operator and the franchising authority are seeking special authority to undertake an experimental approach, community groups have an interest in seeing that the argument in favor of authorizing the experiment is put forcefully and well. The community should ensure that the application carefully details how an experimental approach will be fulfilled. Some communities with a zeal for innovation have found themselves with systems that do not operate at all. In some cases, a prospective franchisee might have agreed to a community-imposed clause in the franchise in hopes that the FCC would strike the clause. Such hopes might take the form of a weak defense of the clause in the certificate application. This problem may arise where a franchise fee in excess of the federal standard has been established.

Even when the community has done an excellent job of drafting a cable franchise, the application for a certificate of compliance should be considered with care. A franchise document will usually be limited to the necessities, the formal terms of the franchisee's duties, but it will be surrounded by the community's tacit understandings and expectations. The certificate application is an opportunity to address those understandings specifically.

Once the group has analyzed the document, its comments to the FCC do not have to take any precise form. Of course, a certain formality of presentation and thoroughness of analysis is likely to dignify the group's views. If possible, the group should also document any claims it makes in its comments—for example, if it is claiming that the application fails to provide adequately for educational needs. And statements should be supported by appropriate affidavits where possible. The services of a lawyer are advisable, but not indispensable, in drafting comments on a certificate of compliance application.

For a sound technical reason, it is important to file comments or objections to a certificate of compliance application. Once an application is being processed, it can be amended. Groups that have filed objections and are therefore "parties" are entitled under Sec. 76.18 of the rule, as a matter of right, to notice of these amendments and the opportunity to comment on them. Moreover, it may not be possible, at a later date, to raise objections that were not presented at the initial certification stage.

VI. AN ANALYTIC APPROACH TO CERTIFICATE APPLICATIONS

There are several ways to analyze an application. Special interest groups, such as education users or public access advocates, will often be interested in specific aspects of the application that touch upon their concerns. A checklist approach can be used, comparing the application with the goals developed by the group in the franchising process. This section suggests the kind of questions to be asked in analyzing the application.

ARE THERE VIOLATIONS OF FEDERAL STANDARDS?

The clearest function of the reviewer is to determine whether the application reveals or disguises violations of federal standards. For example, the application must state that there was a "full public proceeding affording due process" before the franchise was granted, but a community group may argue there was not. The rule is vague, however; the legal concept of "due process" may mean anything from an informal conference to a trial-type adversary proceeding. But the certification process is a forum for clarifying the test on a case-by-case basis.

The Commission's recent actions, however, indicate that it will not require strict adherence to its franchise standards—at least for grandfathered systems. Its *Reconsideration Opinion and Order* amended the "grandfather" clause of Sec. 76.31(a), stating that cable systems that were franchised but nonoperational before March 31, 1972 need be only in "substantial compliance" with FCC franchise rules. Further, the FCC's first decision under the amended rule indicated that "substantial" can be rather minimal. In disposing of a combined certification and special relief application, the Commission approved a franchise that provided for a sliding fee scale of 5 to 12 percent, contained no construction requirements, ran for twenty years with automatic renewals, and had no subscriber complaint procedures—thus effectively violating every single provision of Sec. 76.31(a). If such actions are to become commonplace, the Commission was using understatement when it announced that "henceforth, the term 'substantial compliance' will be given liberal construction." Though the FCC attempted to confine its decision to grandfathered systems, its actions suggest that it will enforce the franchise standards loosely.

This attitude cuts both ways for citizen groups. On the one hand, the Commission's laxity may allow cities to push through franchises without public participation or consumer protection. On the other, the Commission's liberality may permit citizen groups to sustain franchises that embody generous commitments of fees and channels.

A community group may object to a franchise's definition of "significant construction" or "equitable and reasonable extension," pursuant to Sec. 76.31(a)(2). These terms, too are vague. Nevertheless, a franchise that allowed a cable operator more than a year after certification to complete his headend or allowed him to "cream skim" the more affluent areas of his franchise area presumably would violate the rule. This type of violation will be difficult to prove, however. No responsible city or cable attorney will allow such terms in the franchise; instead, they will result from informal understandings between the cable operator and the city government. The community may wish the certificate application to be far more detailed than the franchise itself with respect to the construction timetable, perhaps listing specific neighborhoods with target dates for cabling. Even so, a pattern of delaying tactics or selective wiring may become evident only after the system has been in operation for some time. As a result, a citizens group may need to file subsequent petitions under Sec. 76.7's general relief provisions.

A community organization may argue that the term of an initial franchise is in effect longer than fifteen years, or that a renewal franchise is not of "reasonable duration" under Sec. 76.31(a)(3). The proof of the first argument will depend on how automatic the right of renewal is. A community group might urge the FCC to impose conditions that will transform the renewal process into a meaningful opportunity for local review. And where a franchise sets a definite renewal period, of course, a local group may argue that the renewal duration is not "reasonable." Once again, only precedent will give the "reasonable" standard any real meaning; but the FCC appears to be taking a very lenient attitude in applying the term to grandfathered systems.

A citizen group may object under Sec. 76.31(a)(4) that a franchising authority has not "specified or approved" initial subscriber rates or that a franchise does not provide for "an appropriate public proceeding affording due process" before rates may be changed. Where the franchise and the application merely parrot the language of the regulations, the group may argue that the franchisee and the franchising authority must give a more detailed notion of the procedures and standards to be used when rate increases are to be considered.

A citizen group may argue that a franchise violates Sec. 76.31(a)(5) by failing to "specify procedures" for subscriber complaints or by not requiring the cable operator to "maintain a local business office or agent." Since the Commission obviously has not defined adequate "procedures," the group presumably must bear the burden of proof that a particular method is inadequate. Since the cable system often will not be in operation when the operator files his application, it may be difficult to measure the adequacy of the procedures. Nevertheless, some situations may create almost a presumption of inadequacy. For example, location of a cable system's main office far from a city's core certainly does not presage efficient treatment of central-city subscriber complaints. Here again, though, the FCC does not contemplate close scrutiny, at least for grandfathered systems.

Creative community participation in the certification process should go beyond objections to violations of the rules. The process will enable the FCC to clarify and specify minimum standards for acceptable franchise performance. Citizen participation is essential in this case-by-case definition and amplification of FCC policy. A few examples will show possible directions.

First, the rules do not set a minimum channel capacity for cable systems in smaller markets—that is, those below the top 100. As a result, cable operators in rural areas and small towns may build systems with an inadequate number of channels. Federal standards now are lacking, and the FCC has noted the availability of certification proceedings for measuring the adequacy of a particular channel capacity. Community groups therefore may object to the channel capacity a cable operator promises in his application. Again, objections must be very well reasoned and documented if they are to be successful.

The present rules do not indicate how the transfer of a franchise affects a certificate of compliance. Cable franchises are attractive properties for which there is a brisk market. Transfer of a franchise is particularly important to communities that value local or minority ownership and are concerned with high-quality program origination. Community groups may argue that the certificate of compliance should specify the method of franchise transfer and its effect upon the certificate. For example, a community group might request that a franchise transfer invalidate the certificate, thus forcing the new cable operator to reapply.

A very important opportunity for citizen intervention relates to the access channels mandated by the FCC. The franchise may provide that the cable franchisee furnish three access channels (one each for government, education, and public access) as required by law (in the top 100 markets). The franchise may go on to provide that the franchisee will meet his obligation by providing only these three channels for the entire franchise area even though it may comprise many communities. Indeed, the FCC has indicated that it will approve, on an ad hoc basis, the sharing of access channels by several communities where there is a cooperative franchising process. But there is no set of standards to indicate when access channels should and should not be shared. In New York City, for example, should Manhattan's right to three or more public access channels of its own depend on the pattern by which the franchise is awarded? In many such cases, community groups will wish to argue that the FCC should disapprove sharing in the certification process. Again, the group will have to press its arguments carefully and with documentation. It may argue not only that providing all three channels will not burden the cable operator, but also that the community needs all three channels for itself.

SUPPORTING FRANCHISE PROVISIONS THAT EXCEED FEDERAL STANDARDS

Community groups should not feel that their only role in the certification procedure is to oppose and object. On the contrary, they should support franchise provisions that are more creative and expansive than the federal standards and that serve the community. The FCC has created a difficult balance, encouraging experimentation by franchise authorities but discouraging burdens on cable's growth. When a franchising authority has taken creative action, community groups should underscore its importance before the FCC. Community groups not only must fight for an excellent franchise, but also must defend their product during certification proceedings. Although many franchise provisions may create difficulties with the FCC, two types probably will be the most troublesome: requirements for additional free community channels, and high franchise fees.

Additional Community Channels

The FCC's new rules require major market cable systems to make available one "education access channel," one "local government access channel," and one "public access channel" free of charge. Section 76.251(a)(iv) provides that "no local agency shall prescribe any other rules concerning the number or manner of operation of access channels" without the FCC's special approval. If community groups have been able to secure franchises that require additional channels for free or for preferential rates, they must defend those franchise provisions in the FCC's certification proceedings. This will require a showing that the channels are not only economically feasible for the cable operator, but also necessary for the community.

For example, many observers contend that several channels—not one—are necessary for a truly workable experiment in cable's educational uses. A community with a university or college may need an additional channel for providing higher education to the general public. Some Model Cities programs or community action agencies may want their own channels. A local law enforcement agency may wish to experiment with using a channel for surveillance, information retrieval, or other purposes. All these potential users, of course, presumably can buy time on the leased channels, but the cost may be beyond the means of many grass root community organizations.

If a franchising authority has been able to secure additional channels, it must prepare a persuasive case for the FCC. And the support of community groups can be important and helpful. The local government and citizen groups must present evidence that the additional free or preferentially priced channels are a necessity for the community and not the result of a bartering contest. As noted before, however, the FCC may apply its rule of "liberal construction" to franchise provisions. But since these issues have not yet come before the FCC, the type of evidence necessary is less than clear. Nevertheless, some general suggestions are in order. The local government and community group should attempt to document both present and potential uses of the additional channels. First, they should secure affidavits from any groups that have used the channels, if the system is already in operation. Second, a survey of the existing media's availability to community organizations might be helpful to show that a new medium is needed; affidavits from community organizations or spokesmen could supply the necessary documentation. Third, the local government and community group should analyze and list potential programming sources within the community—such as educational institutions with audiovisual departments, community action programs with videotape facilities, or independent production centers. They could distribute questionnaires to all local individuals or groups with a potential need for channels; these might include such organizations as local charities, hospitals, legal services programs, and welfare organizations. Finally, they must demonstrate why an experimental approach involving reserved channels is preferable to a scheme where channels are possibly available for leasing.

Both local governments and community groups may argue that currently grandfathered systems should provide more channels than the FCC requires. Under the present rules, any cable system in operation before March 31, 1972 must comply with the access requirements only to the extent that it carries additional broadcast signals. Moreover, the rules specifically state that for each additional broadcast signal the system must add only one channel and only "in the following order of priority: (1) public access, (2) educational access, (3) local government access, and (4) leased access."[1] Thus, a local government or community group may request that the FCC require a cable system to add all the channels or that they be added in a different priority. They might make the type of showing necessary to justify more channels than the rules require. Or they might argue that the cable system has constructed so little that it does not merit grandfathering. All of these requests and arguments face an uphill battle.

In a release from the FCC Cable Television Bureau, the Chief of the Bureau attempted to clarify some frequently raised issues concerning access channels. For example, as indicated in this report, many communities may wish to have more access channels than the cable rules permit (without FCC ratification). The Bureau letter states that such additional channels will not be permitted "unless during the certificating process the Commission is shown that such additional channels are necessary and capable of being used according to an existing, viable plan."

While this opinion is not binding, it should be heeded. It underscores the need for the franchising authority and community groups to have documented plans to submit with the application for a certificate. Of course it is hard to tell what is "necessary" and what is "viable." The plan should probably indicate why other channels of communication are inadequate; it should also stress that agencies that say they will use the cable access channels have the capability to do so.

The Bureau letter also states that a franchise will be disapproved if it requires all access services to be made available at no charge. This may mean, for example, that the requirement in the rules that 5 minutes of production time can be provided free of charge on the public access channel cannot be greatly exceeded. But the Bureau letter softens the impact of its statement by saying that free or reduced-rate services can be made available on an experimental basis.

Finally, the Bureau letter states that franchising authorities outside the 100 major television markets can require access services, but "to no greater extent then the Commission requires for systems in major markets." It is difficult to know what this Bureau ukase means. It may mean that smaller communities that require those access services which the FCC imposes on large systems can do so without special justification before the FCC in the certification process. Or it may mean that even if a justification is put forward, no small-market franchise will be certificated if it goes beyond the rules for major markets.

---

[1] Section 76.251(c). The FCC thus retreated from its previous position that a grandfathered system would have to add all the designated channels if it added any broadcast signals. See *Reconsideration of Report and Order* at 13867.

Excessive Franchise Fees

The FCC prohibits the franchising authority from imposing a higher fee than 3 percent of a system's gross annual subscriber revenues without special FCC approval. To justify a higher franchise fee, the franchising authority and cable operator must make a joint special showing that the fee "will not interfere with the effectuation of federal regulatory goals" (the burden is on the franchisee) and "is appropriate in light of the planned local regulatory program" (the burden is on the franchising authority). (Section 76.13(b).) Thus, the rules raise two separate issues: first, what amounts are included in the franchise fee, and second, what is a "local regulatory program?" The views of community groups on these issues will be important during the certification process, since the FCC appears to be taking a lenient view of its own standard.

First, a franchise fee obviously includes more than the mere annual payment specified in the franchise; indeed, Sec. 76.31(d) specifically includes "all forms of consideration, such as initial lump sum payments." As a result, the issue is whether a monetary or nonmonetary benefit to the community constitutes a "form of consideration." Since the Commission has yet to pass on these issues, the definition of "consideration" is unclear. Nevertheless, a few potential troublespots are evident. The FCC might hold that additional channels should be included in a franchise fee. Also, free or low-cost community production facilities might constitute a "form of consideration." And a cable operator's direct payment to either a local government or a community group for program production might be included in the franchise fee.

The FCC's Advisory Committee on Federal/State-Local Relations has highlighted ways in which the certification process can insure full disclosure of important financial details of the franchise process. The factors to be taken into account should include:

> Whether a successful franchise applicant was required in a public proceeding to enumerate in writing all of its franchising expenses, preferential equity arrangements, monetary and service grants to state/local institutions and organizations, and service commitments either volunteered in excess of commission maxima or required or agreed to where there are no such maxima.[2]

Second, the Commission has given little guidance in defining a "local regulatory program." Thus, a franchising authority may be unable to justify a higher franchise fee for monitoring the use of a public access channel, for financing an experimental use of channels, or for funding public access users. This restrictive view would cut local regulatory purposes to the quick—i.e., disposition of subscriber complaints, administration of the franchise, and approval of rate changes. The FCC has tentatively indicated a flexible approach on this issue, and its attitude toward franchise fees seems generally benevolent.

These problems may make a community group's support of a higher franchise fee very important. A petition to exceed the 3-percent figure must be supported by the franchising authority and the cable operator. Cable operators often agree grudg-

---

[2] Quoted in *Broadcasting,* November 27, 1972, p. 50.

ingly to high franchise fees, possibly in the hope that the FCC will invalidate them. As a result, a cable operator may not participate forcefully in certification proceedings. Community group support may become increasingly important.

Despite the FCC's apparent liberality, local governments and community groups should demonstrate the need for additional franchise revenue. The franchising authority should identify and emphasize all costs related to its regulation in any way. These could include expenses incurred during the franchising process in relation to contracts with independent consultants, conduct of public hearings, publication of notice, and the like. The franchising authority can either amortize these costs over the franchise period or collect them in a lump-sum payment. The franchising authority should attempt to characterize its actions as "regulatory." For example, it might use franchise receipts initially for programming to inform subscribers of the opportunities for access and rights to service. Community groups could argue that a "regulatory purpose" includes promoting public access and other nonbroadcast uses by furnishing funds for production uses. Finally, the franchise fee might be structured imaginatively. For example, the franchise might provide for an increase in the fee if the number of subscribers reached a particular level—e.g., 10,000 —since the FCC recently spoke of such an arrangement with approval. Or the fee might be higher for nonsubscriber than subscriber services, such as subscription television, home safety monitoring services, and shopping by cable. Since the rule speaks in terms of "the franchisee's gross subscriber revenues per year," a higher fee for revenues derived from cable users—as opposed to cable subscribers—might be valid. The Cable Television Bureau has stated, however, that a franchising authority may not "impose a franchise fee based upon revenues derived from 'auxiliary' services such as advertising revenues, leased channel revenues, [or] pay cable revenues. . ." But this opinion has not been tested before the Commission itself. Undoubtedly some communities, wishing to divert the regulatory burden from subscriber revenues, will test this statement of the Cable Bureau's policy.

SHOULD THE FCC ADD NEW REQUIREMENTS?

A franchising authority sometimes will adopt a franchise that is valid under federal requirements but unacceptable to community groups. This situation raises difficult questions for both community groups and the FCC. For example, community groups may develop an excellent experimental educational program with the full support of local educational authorities and a promise of financial aid from the federal Department of Health, Education and Welfare; but the franchise may not include the necessary provisions—sufficient channels, interconnection of schools and homes, and the like. Though not required by federal rules, these provisions may be extremely sensible for the community.

A community group certainly may oppose a certification application on the grounds that it lacks these provisions. The FCC will probably be ill disposed to demands that the franchising authority did not grant. Documentation of tangible local needs and grass roots political pressure may carry some weight with the Commission. If nothing else, this type of opposition may yield negotiated amendment of the franchise or of the certification application, so as to avoid a long and costly administrative hearing.

A SPECIAL CASE: THE CERTIFICATE APPLICATION AND BROADCAST SIGNALS

In the application for a certificate, the cable operator must designate which broadcast signals he plans to carry. Most of the sound and fury on the federal level has centered on cable's use of broadcast television signals. The conflict among cable, copyright, and broadcast interests has produced a number of arcane terms—e.g., "distant signals,"[3] "exclusivity,"[4] "leapfrogging,"[5]—whose legal complexities understandably may intimidate people; but community groups should press on fearlessly through the labyrinth. A cable system's choice of broadcast signals will significantly affect not only local television stations, but also the diversity of locally available programming.

The FCC has indicated that a franchise may not specify what signals a cable system may carry. Nevertheless, objections to the choice of signals can be made during the application process, or petitions for special relief can be filed. But regardless of the vehicle, the local government or community group should consider the substantive areas in which it may wish to intervene.

First, a local government or community group may disagree with the cable operator's choices among geographic sources of distant signals. As noted before, the new rules permit cable systems to import more distant signals than before; at the same time, importation is subject to the "exclusivity" and "leapfrogging" rules. Nevertheless, the cable operator still retains some freedom of choice. Conventional wisdom would allow the cable operator alone to select distant signals because theoretically he will choose the signals most likely to attract subscribers. In actuality, however, he sometimes may not. Importing distant signals usually requires microwave relay systems, whose cost depends on routes and distance. A cable operator therefore may have a very real financial disincentive against importing a signal that many viewers would find attractive. Moreover, he may select signals that do not meet local minority programming needs; and the community group may wish to encourage the importation of television stations that have good minority hiring policies or carry excellent children's programs.

Second, a franchising authority or community group may want some control over the number and type of foreign language stations the cable system carries. The new FCC rules allow unlimited importation of foreign language stations, but the operator may not import the quantity or quality of foreign language signals the community desires; the cost of microwave relays, as well as the potential income from channel leasing, may discourage extensive importation. Conversely, a com-

[3] A television station is considered "distant" if it is assigned to another community or city and cannot be generally received in the cable community without the assistance of the community antenna. The FCC's rules basically determine the quantity of distant signals that can be brought in and the conditions for their use.

[4] The station and copyright owners prevailed upon the FCC to protect existing contract rights. Thus, even if a cable company is entitled to bring in a distant station, it must blackout a motion picture, say, if a local station has exclusivity rights to televise it for a period of years. The exclusivity rules vary with the size of the market. See Rivkin, op. cit.

[5] To strengthen UHF stations and to prevent the growth of national independent stations, the FCC placed some limits on the distant television stations a cable operator could select. These rules, which favor stations nearby or in the same state, are called "leapfrogging" rules because they try to prevent excessive leapfrogging by cable systems. See Rivkin, op. cit.

munity might worry that importation would undermine the economic viability of a local foreign language station.

Third, a local government or community group has a similar interest in importation of educational television stations. As noted in Sec. III, the new rules allow a cable operator to import an unlimited number of educational signals, although a local educational station may object if it can document substantial economic harm from importation. In any event, a local government or community group may find itself torn by conflicting desires. On the one hand, it may fear that importation will kill a budding local educational station. On the other hand, the cable operator may not import enough educational signals because of microwave costs and potential leased channel revenues.

Fourth, a local government or community group may object to the programming a cable operator substitutes for shows "blacked out" by the exclusivity rules. This may create several difficulties for the community. Since blacking out often will be necessary in prime time and on popular channels, the freed time slots will be valuable. Community groups may wish to use this time for local programming rather than for imported or "canned" shows. For example, a group might wish to use a time slot for a college extension course or for public access programming. In addition, the rules do not subject substituted programming to the "fairness doctrine." As a result, a substituted program might present only one side of an issue. Community groups and local governments therefore may wish to exercise some form of fairness control over substituted programming.

Finally, a local government or community group may be concerned with cable's economic impact on local television stations. Even the severest critics of broadcast television concede that it does some things well: it serves both the rich and poor, it carries events of national significance, and it entertains. As the Commission's six-year delay in lifting the cable television freeze amply indicates, it is unsure about cable's impact on broadcasting. This continuing uncertainty may force local groups to take seemingly contradictory positions. On the one hand, a local government or community group might join a television station in petitioning the FCC for special protection on the grounds that the cable system is decreasing the station's revenues and therefore its public service capability. On the other hand, a community organization might oppose a television station's petition for special relief on the grounds that its revenues are not low enough to justify neglect of public service programming. A local government or community group therefore can help to distinguish the valid from the invalid in a local station's claim of economic injury.

WHAT REMEDIES CAN THE COMMUNITY GROUP SEEK?

As noted in Sec. II, denial of a certificate may have no legal effect on the validity of a franchise; instead, it may only prevent a cable operator from carrying broadcast signals. Realistically, nevertheless, denial of a certificate will abort a system, since it still needs broadcast signals to be economically viable. The franchise should therefore state explicitly the effect of the FCC's refusal to grant a certificate. A number of options are open to the local franchising authority. First, the franchise might provide that it will remain in effect, except for terms invalidated by the FCC.

Second, the franchise might require that the selection process be reopened completely. Finally, and most realistically, it simply might empower the franchising authority to negotiate new terms acceptable to the FCC.

If a franchising authority adopts this last alternative, it should include a reservation of power clause in order to retain flexibility in negotiations. It might go even further, however, and include franchise provisions to replace those invalidated by the FCC. For example, a franchise might provide that if the fee is invalidated, the cable operator must complete construction of his trunk line more quickly. The very existence of alternative provisions, however, may function as a double-edged sword. On the one hand, the FCC's recent actions indicate that it probably will follow a clear manifestation of local desires in fashioning a remedy for denial of a certificate. On the other hand, the presence of an alternative may be an invitation to invalidation; it not only undercuts the franchising authority's argument that the term is essential, but also ensures that denial of a certificate will create only minimal disruption.

Where alternative franchise provisions do not exist, the local government and community group should suggest alternative remedies to the FCC. Denial of certification often will result in further negotiations among the cable operator, the franchising authority, the community group, and perhaps even the FCC's staff. The community group should have a list of alternative approaches it would find satisfactory where it appears that it has successfully objected to a franchise provision. Often these discussions with FCC staff members can take place before actual denial.

VII. SPECIAL WAIVER AND RULEMAKING PROCEDURES

After the system is in operation it may seek relief from various FCC or franchise requirements. There are two principal ways of obtaining relief. First, an organization may petition the FCC to waive a particular rule's application to it. As noted in Sec. II above, Sec. 76.7 of the FCC rules grants an explicit, albeit limited, right for any "interested person" to request waiver or modification of any cable television rule. Second, anyone may request the Commission to make a new rule or to change an existing rule, by way of a petition for proposed rule making. Cable operators, broadcasting stations, program producers, equipment manufacturers, and other interest groups besiege the Commission constantly with waiver or rulemaking petitions. Local governments and community groups, however, hardly ever use either procedural technique and usually are unprepared even to file comments on petitions that affect them.

REACTION OF LOCAL GOVERNMENTS AND COMMUNITY GROUPS TO PETITIONS FOR SPECIAL RELIEF

Cable operators and broadcasting stations often seek petitions for special relief. A few illustrations will show the scope of waiver applications and their relation to community interests.

A cable operator may seek a waiver from the requirement that:

- The system originate programming "to a significant extent as a local outlet" if it has more than 3500 subscribers. Section 76.201(A).

- A cable system located outside of all major television markets have facilities available for origination of local programs. Section 76.201(B).

- A major market cable system have a minimum 20-channel capacity and two-way capability. Section 76.251(a)(1)(2).

- The system provide additional channels when existing channels are in relatively high demand. Section 76.251(a)(8).

279

- The system adhere to the technical standards established in Subpart K of the rules.

- The system provide dedicated access, educational, and governmental channels. Section 76.251(a).

- The franchise contain a 15-year limitation on the initial duration or that it require significant construction equitably spread throughout the franchise area. Section 76.31(a).

- A television station not own a cable system within its service area. Section 76.501.

- Existing cable systems come into compliance with the rules within five years.

A local television station may also seek special relief:

- It may petition the FCC to decrease the number of distant signals which a local cable system may import, in order to diminish competition with the cable system.

- A local foreign language station may object to the importation of distant foreign language stations.

- A local educational station may object to the importation of distant educational signals.

- A commercial station may demand that a cable system not duplicate its programs with distant signals, even where the rules allow nonsimultaneous duplication.

- A station may petition for more stringent leapfrogging rules, in order to restrict the sources of a cable system's distant signals.

Many of these petitions will seem technical and irrelevant, but a local government or community group should examine every application in detail to determine its relevance to community interests. As noted in Sec. III, importation of distant signals may affect not only the programming available to subscribers, but also the public service capability of local television stations. And special interests in each community undoubtedly will want to be heard on some issues; for example, parent-teacher associations may have strong views on the education channel's uses.

Unfortunately, however, community groups now have no method of automatically receiving notice about petitions that affect them. Section 76.7(b) provides that a petition for special relief must be served upon the franchising authority or any "interested person who may be directly affected if the relief requested should be granted." But this may not necessarily include community groups that represent only a general public interest. A community group might send a formal letter to a cable operator stating that it is an "interested person" in relation to any petitions for special relief. It is not certain whether this tactic would create any legal right to notice. Nevertheless, it may have some impact on the cable operator—especially if copies are sent to the franchising authority and to the FCC.

What are the requirements to oppose a petition? Section 76.7(b) explicitly states that a "petition may be submitted informally, by letter." Accordingly, any comments or opposition presumably also may be informal. As a result, community groups should not be deterred from filing oppositions because they lack an attorney. As with an opposition to a certification application, the document should be as detailed as possible and present a well-documented case. As with an opposition to a certification application, local governments or community groups should point out inaccuracies or omissions in the cable operator's statements. For example, a cable operator may seek a waiver of the public access channel requirement on the grounds that the channel has generated little use or interest. A community group might oppose this by showing that the cable operator has not provided adequate studio facilities, that studio facilities have been located inconveniently, or that the cable operator has exercised censorship. Finally, Sec. 76.7(c)(2) provides that the cable operator's petition must state "all steps taken ... to resolve the problem." A community group or local government might argue that the cable operator has failed to take readily available steps. For example, a cable operator may seek a waiver of the educational channel requirement, on the grounds that the local school district has not used the channel. A community group might file affidavits showing that other educational institutions—such as a state university or a parochial school—never have had an opportunity to use the channel.

Preparation of an effective opposition sometimes may require fairly sophisticated economic analysis—particularly where a local broadcast station argues that a cable system creates unfair competition, or where the cable operator argues that his subscriber base will not support required services. Though cost analyses seem difficult and mystifying at first, most community groups should seek enough expert advice to rebut or at least analyze the petition.

The presence of a local voice is highly important for several reasons. The FCC usually hears only from one side when it passes on a waiver application. Moreover, the Commission's staff is neither large nor expert enough to check out the accuracy of all statements. Finally, the FCC often grants certain categories of waivers automatically if there is no opposition. If no one speaks for the community, the FCC may act on the principle that silence gives consent.

PETITIONS FOR SPECIAL RELIEF BY LOCAL GOVERNMENTS AND COMMUNITY GROUPS

A local government and community groups can seek additional or different requirements for a cable system by opposing a certification application. They also have recourse to Sec. 76.7(a), which creates an avenue for special relief not solely for the cable operator but for any "interested person."

Community groups thus may come before the FCC on their own. Recent federal court and FCC rulings indicate that the term "interested person" includes local viewers and subscribers. The FCC may not actively encourage public intervention, but the rules undeniably permit it. In recent years many groups have intervened to challenge license renewals of broadcast stations.

Community groups might use Sec. 76.7 petitions in many ways. As discussed above, a petition for special relief is a possible—albeit not probable—way of securing provisions that the franchise omits. Community groups also may use petitions to protest inadequate performance, just as they may use oppositions to certification applications to protest inadequate promises. For example, a community group might argue that a cable operator is not fulfilling his obligation under Section 76.201(a) to operate "to a significant extent as a local outlet by origination cablecasting." The meaning of the requirement is vague, and a community group could secure an interpretation through a petition. In short, petitions for special relief allow community groups to attack the actual—as well as the promised—performance of their cable systems.

This avenue may be partially closed off in the future, however. The FCC recently adopted an amendment that will limit recourse to Sec. 76.7 by barring claims for relief that could have been filed under Sec. 76.27 but were not. As noted before, objections to certificates of compliance have to be filed within 30 days. There is no such limit on 76.7 applications for relief. The FCC seemed worried that citizen groups and others would use Sec. 76.7 to obtain relief they could no longer obtain under the certification objection method. The new rule states:

(i) If the relief requested could have been earlier filed pursuant to § 76.27, the petition will be dismissed unless the petitioner shows that:
  (a) The facts relied on relate to events which have occurred or circumstances which have changed since the last opportunity to present such matters pursuant to § 76.27.
  (b) The facts relied on were unknown to petitioner until after his last opportunity to present such matters, and he could not through the exercise of ordinary diligence have learned of the facts in question prior to such opportunity.
  (c) Consideration of the facts relied on is required in the public interest.

While the motivation for the new rule is clear enough, its terminology is so vague that it seems a feeble mechanism for discouraging ill-grounded petitions. This is not to say that there is much hope for such petitions even if they receive the FCC's attention. The rule's reference to new facts or changed circumstances in Subsec. (a) probably will require a petitioner to prove there has been a massive transformation of the local situation, if the phrase is interpreted in the light of past judicial and administrative constructions of similar provisions. Only extreme changes may suffice, such as the discovery of fraud or corruption on the city's part. And while Subsec. (b) appears to be somewhat more generous, it may not include a community's general ignorance as to the stakes in cable franchising—precisely the situation, of course, that leads to most difficulties. Finally, Subsec. (c)'s reference to the "public interest" is so vague as to be effectively meaningless.

LOCAL GOVERNMENT AND COMMUNITY GROUP USE OF RULEMAKING PROCEEDINGS

Under federal law anyone may petition the FCC to make a new rule or to modify an existing one. Moreover, this right may be exercised in a remarkably easy way. The rules require only that a rulemaking petition be typed in any one of several simple formats and that it state the reasons for adoption or modification of the rule. Industry representatives file hundreds of rulemaking petitions each year; representatives of the "public interest" file virtually none.

Rulemaking petitions are only occasionally effective. A petition generally will not goad the Commission into adopting a rule, but at least may persuade it to consider a new policy issue, in which case new rules may follow later. For example, in 1967 the Office of Communication of the United Church of Christ filed a short petition requesting the Commission to ban racial discrimination by broadcasters. The Commission responded by first adopting a general policy against discrimination and then promulgating a comprehensive set of rules *(Nondiscrimination in Employment Practices,* 37 Fed. Reg. 6586, March 31, 1972).

SOURCES OF AID FOR LOCAL GOVERNMENTS AND COMMUNITY GROUPS IN SPECIAL RELIEF AND RULEMAKING PROCEDURES

The rules thus give local governments and community groups at least an opportunity to participate in FCC proceedings. This participation need not be as difficult as the rules' mystifying language might indicate. Nevertheless, both local governments and community groups obviously can use help. A number of sources exist now, and others will become available.

The Office of Communication of the United Church of Christ works through its local affiliates to advise communities. The American Civil Liberties Union monitors FCC matters that touch on First Amendment interests. The new Cable Television Information Center will provide information to local groups and will contract to do studies for local governments. In some communities, publicly supported legal services offices or "public interest" law firms[1] will help groups file with the FCC; though most attorneys in these offices have little experience in communications law, they are often willing and able to contribute their time. In addition, regional and national organizations dealing solely with cable have developed and will become increasingly useful. Finally, some cities, such as New York, attempt to monitor issues which may affect local interests.

Before a local government or community group can take meaningful action, it must know that an issue will affect it. For example, another community's petition for special relief may be relevant, since the FCC's disposition will become a precedent. As a result, coalitions of community groups and local government should file *amicus curiae*— "friend of the court"—comments when a seemingly unrelated issue may ultimately affect them. Some communities and organizations have picked a person or committee to review relevant literature—such as magazines, newsletters, and trade publications.

Finally, a local government or community group always should attempt to settle a dispute. Negotiation is obviously much cheaper and easier than litigation for all parties. Nevertheless, community groups often will find that a cable operator is much more amenable to settlement if the group already has filed an intelligent and persuasive document with the FCC.

---

[1] A list of such firms can be obtained by writing to the Center for Law and Social Policy, 1600 20th Street, N.W., Washington, D.C. 20009. For an extensive list of information sources regarding cable, see Baer's compilation under the heading "For More Information" at the end of *Cable Television: A Handbook for Decisionmaking,* Crane, Russak & Co., New York, 1974.

VIII. CONCLUSION

Local governments and community groups can participate in the FCC's certification, waiver, and rulemaking proceedings. Citizen participation inevitably will create conflicts among the FCC, local governments, community groups, and cable operators. This price, however, is well worth paying. First, the heat of battle may refine the FCC's policy decisions. Second, and more important, local governments and community groups must act immediately and affirmatively to prevent cable regulation from becoming an exclusively federal domain. Citizen participation may inspire a truly creative dual federalism. For it is clear that in many cases, effectuation of federal goals will be accomplished only if strong voices are heard from the local level.

Index

Access, channels, 125-128, 262, 271, 272-273
  investigating uses of, 79 80
Aerial versus underground installation, 9-10, 104-105, 184-185
Ameco DISCADE system, 22
American Civil Liberties Union, 97, 100, 107, 112, 131, 132, 139, 283
American Television & Communications, 50
Amplifiers, 10-11, 169
  bridging, 11, 183
  cascading of, 10-11, 176, 181
  and channel capacity, 15
  line-extender, 11, 184
  as source of interference, 60, 181
  for two-way transmission, 43
Antennas, 5
  as source of interference, 179
  of subscribers, maintenance of, 141-142
Applicants for franchise
  background information on, 85-86
  chart for evaluation of, 89
  decision on selection of, 87-90
  fees for, 85, 132-135, 263, 274-275
  financial qualifications of, 88
Applications for franchise, analysis by community groups, 267-278
Attenuation of signals, 4, 10, 181
Audience. See Subscribers
Audiotronics equipment, 27
Automated services, 7, 9

Bandwidth, 4, 15
Bids and awards, in franchising, 74-75
Billing and payment procedures, 119
Bonds, performance, 137-138
Boundaries for franchise areas. See Geographic boundaries
Bridging amplifiers, 11, 183
Broadcast signals, classes of, 4
Broadcasting, compared to cable television, 4-5, 9
Business data transmission, 56-57, 203-204

Cable television
  communication capacity of, 15-24
  compared to broadcasting, 4-5, 9
  components of, 5-11
  new services with, 49-57

services offered by, 14
technical standards for, 58-67
Cable Television Information Center, 96, 105, 108, 283
Cable Television Relay Service, 35, 186
Cablecom-General, 50
California, cable systems in, 47, 50, 194, 204
  in Beverly Hills, 100, 103, 104, 105, 114-115, 116, 117, 118-119, 120, 121, 125, 127, 134
  Kern Cable Company, 9-10, 118
  in Los Angeles, 123, 171
  in Redlands, 167, 221-223
  in San Mateo, 105, 135, 137
Cameras, 27-30
Cancellation of franchise, 143-145
Capabilities of various cable systems, 205
Cascading of amplifiers, 10-11, 176, 181
Cassette tape recorders, 30, 54
Certificate of compliance, 90-91, 259, 264-266
  analysis of applications for, 267-278
Channel capacity, 15-24, 163-164, 169-174, 261, 271
  basic 12-channel system, 15-17, 170
  expanding of, 17-22
  FCC regulations on, 3, 13, 15, 130, 172, 246, 261
  minimum, 130-131
  nonstandard channels, 21
  and receiver design, 23
Channels
  allocation of, 125-128
  classes of, 58, 61, 165-166, 217
  local control of, 154-155
Chicago ordinance, 86
Cinca Communications, 50
Citizen participation in planning, 76, 229-256
  in ad hoc committees, 253-254
  and access channels, 271, 272-273
  and allocation of channel uses, 246
  and alternative remedies offered, 277-278
  and analysis of applications, 267-278
  and assessment of community needs, 76-81
  and choice of broadcast signals, 260-261, 276-277
  in conferences, 253

in delegated groups, 254
early developments in, 233-236
federal programs affecting, 236-239, 249-250
forms of, 249-254
and geographic size of system, 243-244
initial steps in, 248-249
issues involved in, 239-241
and local control of channels, 154-155
and monitoring of performance, 246-247
and ownership of system, 242-243
and petitions for special relief, 281-282
recommendations for, 255-256
and requirements of FCC, 275
sources of aid for, 283
and special waiver and rulemaking procedures, 279-284
and subscription fees, 245-246
and support of provisions exceeding federal standards, 271-272
and surveys, 250-252
and view of municipal officials, 230-232
and violations of federal standards, 269-271
See also Public hearings
Coaxial cable, use of, 5
Common ownership of cable systems, 141
Communication capacity. See Channel capacity
Community Antenna Relay Service, 35, 186
Community needs, assessment of, 76-81, 250-252
See also Citizen participation
Competitive bids and awards, 74-75
Complaints by subscribers, procedures for, 109-111, 270
Compliance. See Certificate of compliance
Computers, in two-way communication, 39
Conferences
and citizen participation, 253
teleconferencing, 55-56, 203
Connecticut, cable systems in, 79
Connections to cable system
disconnection and reconnection charges, 116-119
fees for, 114-116
Construction
costs of, 10, 13, 184
monitoring of, 91
requirements for, 104-107
safety and damage requirements in, 105-107, 138-139
timetable for, 101-102, 155, 263, 270
underground versus overhead, 9-10, 104-105, 184-185
Converters, 11, 16, 19-22, 171-173, 196, 203
costs of, 21-22, 172
interference from, 60, 183
for pay-TV, 49
in switched systems, 22
types of, 20
Costs. See Economic considerations
Council of Governments, 79
Miami Valley, 102, 110, 130, 140, 141
Cox Cable, 50
Cross-ownership of cable systems, 141
Cypress Cable System, 244

Damages, liability for, 105-107, 138-139
Data-response services, 56-57, 203-204
Decisionmaking process, in planning for cable system, 80-81
Decoders, for pay-TV, 49
Demodulation, 7
Demonstration projects, for two-way communication, 44-48
Design of system, 198-206
for basic system, 199, 203
and capabilities of various systems, 205
comparison of options for, 24
constraints in, 196-197
for data-response services, 202, 203-204
and franchise provisions, 206-209, 212-213
for private channel cablecasting, 200, 203
for two-way communications, 201, 203
Disabled persons, reduced rates for, 121
Disconnection charges, 116-119
Distortion and noise. See Interference
Drop cables, 11, 183
Dual cable systems, 17-19

Economic considerations
amplifier and cable costs, 181
billing and payment procedures, 119
camera tube costs, 29
and cancellation of franchise, 143-145
comparison of design options, 24
construction costs, 10, 13, 184
converter costs, 21-22, 172
for data transmission, 57
and economies of scale, 153
fees charged. See Fees
financial qualifications of applicants, 88-89
for frame-stopping terminals, 53
interconnection costs, 35, 187, 188
investment requirements, 155
and liability for damages, 105-107, 138-139
mobile equipment costs, 30
monitoring costs, 64, 168-169
for multiple cable systems, 19
overhead versus underground construction costs, 10, 184
pay-TV programs, 49-52, 128, 203
and performance bonds, 137-138
pole rental rates, 10
production costs, 32
in receivership, 143
for studio facilities, 9, 27, 28
for switched systems, 22
tape recorder costs, 29, 54
for teleconferencing, 56
two-way installation costs, 39, 43, 44, 46, 175
Educational channels, 126-128, 277
Electronic Industrial Engineering equipment, 47
Emergency use of facilities, 142
Employment practices and training, 107-108
Equalization of signals, 7
Equipment
for distribution of programs, 5-11
mobile facilities, 30

for production facilities, 9, 25-27
for reception of programs, 11
Exclusivity rights, 99-100, 261, 276
Expansion of services. *See* New services
Expenses. *See* Economic considerations
Expiration of franchise, 143-145

Facsimile devices, 53-54
FCC rules
  on access channels, 129, 262
  on allocation of channels, 16, 125-128
  background of, 260-263
  on cable capacities, 13
  on cable use of broadcast television signals, 260-261, 276
  on certificate of compliance, 90-91, 259, 264
  on channel capacity, 3, 13, 15, 130, 172, 246, 261
  on classes of channels, 165-166
  on complaints by subscribers, 109
  on concentration of control, 141
  on franchise fees, 132-133, 263, 274
  on interconnection of systems, 131-132
  on local origin programming, 262
  on microwave links, 35
  on performance tests, 61-62
  on rates for new services, 125
  on regulatory roles of local governments, 262-263
  on reporting to authorities, 139
  on signal carriage, 101
  source materials for, 72-73
  on subscriber fees, 112, 245
  on technical standards, 58-59, 64, 65-67, 218-220
  on two-way capability, 3, 13, 37, 174
Federal standards
  exceeding of, support for, 271-272
  violations of, 269-271
Feeder cables, 11, 183
Fees
  for basic and ancillary services, 122
  for cable connection, 114-116
  and citizen participation, 245-246
  disconnection and reconnection charges, 116-119
  establishing and adjusting of, 122
  for franchise applicants, 85, 132-135, 263, 274-275
  monthly service rates, 89, 116-119, 127
  for new services, 124-125
  procedures for increases in, 270
  reduced rates for special classes of users, 120-122
  for relocation, 119
  review and revision of, 122-124
  temporary reduction of, 120
  subscriber rates and charges, 112-114
  See also Economic considerations
Field tests, of two-way cable systems, 44-48
Filters, in two-way transmission, 43, 44, 175, 176
Financial aspects. *See* Economic considerations
Florida, cable systems in, 47, 50
Foreign language stations, 276
Frame-stopping terminals, 52-53

Franchising, 71-155
  and allocation of channels, 125-128
  approaches to, 74-75
  and assessment of community needs, 76-81, 250-252
  cancellation and expiration provisions, 143-145
  and certificate of compliance, 90-91, 259, 264-266
  and chart for evaluation of applicants, 89
  checklist of major elements in, 147-150
  citizen groups in, 76, 229-256
  and concentration of control, 140-141
  and connection fees, 114-116
  and construction requirements, 104-107
  and continuing administration, 91-92
  decision on awarding of, 87-90
  definition of terms in, 94-96
  and design of system, 206-209, 212-213
  and disclosure of applicant's background, 85-86
  and dissemination of request for proposals, 84-86
  duration of franchise, 97-99
  and emergency use of facilities, 142
  and employment practices and training, 107-108
  and explanation of decisions, 82, 84
  and extent of wiring, 102-103
  and fees for applicants, 85, 132-135, 263, 274-275
  and flexibility in decisionmaking, 80-81
  geographic boundaries in, 78-79, 243-244
  and geographic exclusivity, 99-100, 261, 276
  guidelines for, 89-90
  hearings and tentative decisions in, 81-83, 86
  and information on proposed routes, 86
  and interconnection of systems, 131-132
  for joint ventures, 83
  and liability for damages, 105, 107, 138-139
  and maintenance of home antennas, 141-142
  milestones in, 207
  and minimum channel capacity, 130-131
  and monitoring of performance, 91, 214, 221-223
  for municipal systems, 83, 84
  for noncommercial systems, 83, 84
  objectives in, 71-72
  and operational standards, 109-111
  and pay programming, 128
  and penalty provisions, 167
  and performance bonds, 137-138
  planning process in, 210-213
  prefatory provisions in, 93-94
  preparation of documents for, 213
  procedural framework for, 75-76
  and provisions for modifications, 142, 206-209, 213
  range of issues in, 80
  and receivership, 143
  renewal of, 97-99
  and reporting requirements, 139-140
  and rights of franchisors, 140
  signal carriage issue in, 100-101
  and single versus multiple ownership, 78, 151-155

and technical standards, 108-109
terms and conditions in, 93-145
transfer provisions in, 135-137, 271
and use of access channels, 79-80

Geographic boundaries of franchise areas,
78-79, 99-100
and citizen participation, 243-244
and divided ownership, 153
Ghost images, source of, 16, 170
Government, local
and boundaries for franchise areas, 78-79
channels used by, 125-128
petitions for special relief, 281-282
regulatory role of, 262-263
rulemaking petitions by, 282-283
sources of aid for, 283
and teleconferencing, 55
Grandfathered systems, 97, 125, 133, 134,
263, 269, 270, 273
Gridtronics, 50

Headends, 7
in hub systems, 11, 33-34, 181, 244
relays to, 5
as source of interference, 179
Health services, 55
Hertz frequencies, 15
High-resolution TV, 54-55
Home Box Office, 50
Home Theatre Network, 50
Hughes-Theta Subscriber Response System,
204

Illinois
cable systems in, 47
Chicago ordinance, 86
Commerce Commission, 99, 110, 140,
142
Image Orthicon tube, 27, 29
Information retrieval services, 52-53
Installation costs, 114-116
Institutional services, 55-57
two-way communication in, 41
Instructional Television Fixed Service, 188
Interconnection of cable systems, 33-36,
131-132, 185-188, 197
area, 33-34
by cable, 35-36
and franchise boundaries, 78-79
internal, 33
by microwave, 35-36, 186-187
networking, 34-35
and technical compatibility, 151-152
techniques for, 35-36, 186-187
Interference, sources of, 16, 19, 21, 60,
170-172, 176, 177, 179-184, 198
in two-way transmission, 41, 43
Investment requirements, 155

Joint ventures, franchise for, 83

Kansas, cable systems in, 39
Kern Cable Company, 9-10, 118
Kilohertz, 15

Laser Link Corporation, 187
Leapfrogging rules, 261, 276

Leasing
of poles from utilities, 9-10
of receivers by cable operators, 23, 140-
141
Liability for damages, 105-107, 138-139
Line extender amplifiers, 11, 184
Live programming, facilities for, 25-27
Local control of channels, 154-155
Local Distribution Service, 35-36
Local programming, facilities for, 7-9
Louisville ordinance, 84-85, 94-96

Magnavox, 50, 172
Mail, electronic, 53-54
Maintainability of system, 178-185, 197
for home antennas, 141-142
standards for, 61
Massachusetts, cable system in, 55
Megahertz, 15
Microwave transmission, 5, 7, 35, 186-187
Midband channels, 19, 21
Minority groups
employment and training of, 107
participation in planning, 231, 239-240,
242-243, 244
Mixing of signals, 7
Mobile facilities, 30
Modification of franchise, provisions for,
142
Modulation, 7
Monitoring of performance, 61-63, 91,
166-169, 214, 221-223
and citizen participation, 246-247
cost of, 64, 168-169
Monthly service rates, 116-119, 127
Multiple cable systems, 17-19
Multiple-hub cable system, 11, 33-34, 181,
244
Multiple versus single ownership, 151-155,
185
Municipal systems, franchise for, 83, 84

National Association of Educational Broadcasters, 126, 131
National Institute of Municipal Law Officers, 106, 136, 137, 138, 139, 142
Negotiation approach, in franchising, 74-75
Networking, 34-35, 186
New Jersey, cable system in, 99
Nevada, cable system in, 153
New services, 160-161
compatibility of, 191-193
evaluation of factors in, 207, 208
and franchise provisions, 206-209, 213
requirements for, 188-193, 197
technology for, 49-57
New York
franchise agreements in, 50, 55, 97, 99-
100, 102, 103, 106-107, 109-110, 111,
116, 117-118, 119, 124, 125, 127,
132, 133, 134, 136-137, 141, 142,
143-144, 184, 185, 271
Office of Telecommunications, 134, 168,
214, 247
subdistricts in, 33
TelePrompTer franchise in, 93-94, 166,
185-186, 221
Noise and distortion. See Interference
Noncommercial systems, franchise for, 83,
84

North American Philips, 27
Office of Telecommunications, 214-216, 247
Ohio cable systems, 121
  in Akron, 105, 123, 137, 138-139, 140
  in Dayton, 103, 155, 244, 248, 250, 253, 254
Operational standards, 109-111
Optical Systems, 50
Overhead cables, compared to underground construction, 9-10, 104-105, 184-185
Ownership of system, and citizen participation, 242-243

Panasonic equipment, 30
Pay-TV, 49-52, 128, 203
Penalty provisions in franchises, 167
Pennsylvania, cable systems in, 30, 50
Performance
  bonds for, 137-138
  standards for, 109-111
  tests and monitoring of, 61-63, 91, 166-169, 214, 221-223, 246-247
  yardsticks for comparison's of, 155
Petitions for special relief
  by cable operators, 279-281
  by local governments and community groups, 281-282
Picturephone, 56
Planning process, in franchise decisions, 210-213
Plumbicon tube, 27
Poles for cables, leasing of, 9-10
Polling, in two-way communication, 39, 46, 191
Preamplifiers, use of, 5, 7, 179
Private channels, 49-52, 200, 203
Production facilities, 9, 25-27
  cost of, 27, 28, 32
  for live programs, 25
  for public access channels, 129-130
  sharing of, 153-154
  as source of interference, 179
  for taped programs, 27
Public access channels, 125-128
  facilities for, 129-130
Public hearings, 75, 82
  giving notice for, 81
  on proposals, 86
  on rate changes, 124
Public service channels, 125-128

Quality of signals, standards for, 60
Quality of system, factors in, 177-185

Rand Corporation, 98-99, 100, 102, 141
Rates. See Fees
RCA equipment, 27, 50
Receivers
  leasing of, 23, 140-141
  problems with, 23, 183
Receivership, provisions for, 143
Reconnection charges, 116-119
Recorders, tape, 27-30, 54
Rediffusion system, 22
Reliability of performance, 57, 58, 61, 164, 177-185, 196, 198
Renewal of franchises, 97-99, 270
Reporting to authorities, requirements for, 139-140

Request for proposals, dissemination of, 84-86
Rights of franchisors, 140
Rulemaking petitions, 282-283

Safety requirements, 105-107
Satellite systems, use of, 34-35, 187-188
Scientific-Atlanta equipment, 47
Senior citizens, reduced rates for, 121
Service Electric, 50
Service rates, monthly, 116-119, 127
Shadow trunk cable, 24, 44, 174, 177, 198
Sharing of production facilities, 153-154
Signals
  carriage by cable systems, 100-101
  classes of, 4
  distant, relay of, 5
  mixing of, 7
  quality standards for, 60
Sony equipment, 30
Source materials
  for citizen participation, 283
  for FCC documents, 72-73
Special showing procedures, 126, 263, 279-283
Standards for cable TV. See Technical standards
Sterling-Manhattan CATV Company, 93
Storer Broadcasting, 50
Studio facilities. See Production facilities
Sub-band channels, 21
Subdistricting, 33, 186
Subscribers
  complaints of, procedures for, 109-111, 270
  maintenance of antennas for, 141-142
  rates and charges to, 89, 112-114
  taps and drops for, 11, 183
  two-way services for, 37-39
Superband channels, 19, 21
Super-trunk cables, 35, 181
Surveillance, video, 56
Surveys, value of, 77, 250-252
Switched systems, 22, 173

Tape recorders, 27-30, 54
Taped programs, facilities for, 27
Taps, for subscribers, 11, 183
Technical standards, 58-67, 108-109, 165-169, 196, 218-220, 221
  and costs for monitoring, 64
  FCC regulations on, 58-59, 64, 65-67
  and penalties for noncompliance, 62
  and performance monitoring and testing, 61-63
  problem areas in, 63-64
  for reliability and maintainability, 61
  for signal quality, 60
  writing and enforcing of, 59-63
Telecommunications, Office of, 214-216, 247
Teleconferencing, 55-56, 203
Telephone networks, for data transmission, 56
TelePrompTer, 50, 93, 166, 185-186, 221
Television Allocation Study Organization, 60
Testing and monitoring of performance, 61-63, 91, 166-169

Texas, cable systems in, 47
Theatre Vision, 50
Theta Cable, 50
Theta-Com equipment, 47, 50, 186, 187
Times-Mirror CATV, 50
Timetable for construction, 101-102, 155, 263, 270
TOCOM equipment, 47
Towers and antennas, 5
Transfer of franchise, 135-137, 271
Transformers, for subscribers, 11
Translation of signals, 7, 16, 171
  converters for. See Converters
Trans-World Communication, 50
Trunk cables, 9-10
  dual, 17-19
  shadow, 24, 44, 174, 177, 198
  as source of interference, 181
Tubes, pickup, for cameras, 27
Two-way communications, 37-48, 174-177, 196-197
  costs of, 43, 44, 46
  demonstration projects for, 44-48
  design of system for, 201, 203
  FCC regulations for, 3, 13, 37
  filters for, 43, 44, 175, 176

for institutional services, 41
mid-split technique for, 44, 177, 198
subscriber services for, 37-39
subsplit approach to, 43, 175-176
and teleconferencing, 55-56, 203
terminals for, 39, 50
transmission techniques in, 41-44

UHF stations, 4, 16, 170
Underground cables, compared to overhead construction, 9-10, 104-105, 184-185
United Church of Christ, 283
Utility companies, pole rentals from, 9-10

VHF stations, 4, 16, 170
Vidicon tube, 27, 29
Violations of federal standards, 269-271
Virginia, cable systems in, 50, 53
Waiver procedures, 279-282
Warner Communications, 50
Welfare recipients, reduced rates for, 120-121
Wiring
  and access to premises, 111-112
  extent of, 102-103
  of public facilities, 127

SELECTED RAND BOOKS

Bagdikian, Ben H. *The Information Machines: Their Impact on Men and the Media.* New York: Harper and Row, 1971.

Bretz, Rudy. *A Taxonomy of Communication Media.* Englewood Cliffs, N. J.: Educational Technology Publications, 1971.

Bruno, James E. (ed.). *Emerging Issues in Education: Policy Implications for the Schools.* Lexington, Mass.: D. C. Heath and Company, 1972.

Cohen, Bernard, and Jan M. Chaiken. *Police Background Characteristics and Performance.* Lexington, Mass.: D. C. Heath and Company, 1973.

Coleman, James S., and Nancy L. Karweit. *Information Systems and Performance Measures in Schools.* Englewood Cliffs, N. J.: Educational Technology Publications, 1972.

Dalkey, Norman C. (ed.). *Studies in the Quality of Life: Delphi and Decision-making.* Lexington, Mass.: D. C. Heath and Company, 1972.

DeSalvo, Joseph S. (ed.). *Perspectives on Regional Transportation Planning.* Lexington, Mass.: D. C. Heath and Company, 1973.

Downs, Anthony. *Inside Bureaucracy.* Boston, Mass.: Little, Brown and Company, 1967.

Fisher, Gene H. *Cost Considerations in Systems Analysis.* New York: American Elsevier Publishing Co., 1971.

Haggart, Sue A. (ed.). *Program Budgeting for School District Planning.* Englewood Cliffs, N. J.: Educational Technology Publications, 1972.

Harman, Alvin. *The International Computer Industry: Innovation and Comparative Advantage.* Cambridge, Mass.: Harvard University Press, 1971.

Levien, Roger E. (ed.). *The Emerging Technology: Instructional Uses of the Computer in Higher Education.* New York: McGraw-Hill Book Company, 1972.

Meyer, John R., Martin Wohl, and John F. Kain. *The Urban Transportation Problem.* Cambridge, Mass.: Harvard University Press, 1965.

Nelson, Richard R., Merton J. Peck, and Edward D. Kalachek. *Technology, Economic Growth and Public Policy.* Washington, D.C.: The Brookings Institution, 1967.

Novick, David (ed.). *Current Practice in Program Budgeting (PPBS): Analysis and Case Studies Covering Government and Business.* New York: Crane, Russak and Company, Inc., 1973.

Park, Rolla Edward. *The Role of Analysis in Regulatory Decisionmaking.* Lexington, Mass.: D. C. Heath and Company, 1973.

Pascal, Anthony H. (ed.). *Racial Discrimination in Economic Life.* Lexington, Mass.: D. C. Heath and Company, 1972.

Pascal, Anthony H. *Thinking about Cities: New Perspectives on Urban Problems.* Belmont, Calif.: Dickenson Publishing Company, 1970.

Quade, Edward S., and Wayne I. Boucher. *Systems Analysis and Policy Planning: Applications in Defense.* New York: American Elsevier Publishing Company, 1968.

Sharpe, William F. *The Economics of Computers.* New York: Columbia University Press, 1969.

Williams, John D. *The Compleat Strategyst: Being a Primer on the Theory of Games of Strategy.* New York: McGraw-Hill Book Company, 1954.

The Contributors

WALTER S. BAER, a senior analyst at The Rand Corporation, directed the cable television study and edited this series of volumes. His research interests include the development of two-way communications on cable systems, issues of media ownership, and the technical and economic prospects for new communications systems. He has been a communications consultant to the United Nations and to a number of major corporations. He also directs the Aspen Workshop on Uses of the Cable sponsored by the Aspen Program on Communications & Society. Previously he served on the White House science advisory staff and as a White House Fellow with Vice President Hubert Humphrey in 1966-67. He received his B.S. in physics from the California Institute of Technology and a Ph.D. from the University of Wisconsin. He is the author of reports and articles on communications policy issues.

MICHAEL BOTEIN, presently on leave from the University of Georgia, is a consultant to The Rand Corporation. He was admitted to the New York State Bar in 1969 and taught at the Brooklyn Law School before becoming Assistant Professor of Law at the University of Georgia. He holds a B.A. from Wesleyan, a J.D. from Cornell Law School and an LL.M. from Columbia Law School. He is author of numerous articles on cable television regulation.

LELAND L. JOHNSON is Director of the Communications Policy Program at The Rand Corporation. Prior to joining Rand in 1957, he was an instructor in the Economics Department at Yale University where he received his Ph.D. in 1956. In 1967-68 he served as Research Director of the President's Task Force on Communications Policy, dealing with problems of allocating radio spectrum space, regulation of the telephone industry, and the structure of the television industry. He is a member of the Board of Trustees of the International Broadcast Institute; a member of the Telecommunications Panel of the American Society of International Law; and in 1969-70 was a member of the Twentieth Century Fund Task Force on International Satellite Communications. He has also consulted with The Ford Foundation on uses of satellites for television. He has authored numerous reports and articles on communications industries and their regulation.

CARL PILNICK is a consultant to The Rand Corporation. As president of Telecommunications Management Corporation, Los Angeles, California, he recently completed an analysis of telecommunications for the United Nations. His professional interests include evaluation of broadband networks for urban communications; application of telecommunications to medicine, education and public safety; and evaluation of videocassette technology and applications. He has taught at UCLA and is a member of the FCC Cable Television Technical Advisory Panel. He received his B.S. in Engineering from City College of New York and an M.S. from Stevens Institute of Technology.

MONROE PRICE is a consultant to The Rand Corporation. He received his B.A. and LL.B. degrees from Yale University and was Executive Editor of the Yale Law Journal. He served as law clerk to Associate Justice Potter Stewart of the United States Supreme Court, and in 1965 became Special Assistant to Secretary of Labor W. Willard Wirtz. Since 1966 he has been Professor of Law at University of California, Los Angeles. He is the author of *CATV: A Guide to Citizen Action*, as well as reports and articles on communications policy.

ROBERT K. YIN is a research psychologist with The Rand Corporation and serves part time as Assistant Professor of Urban Studies and Planning at the Massachusetts Institute of Technology. His primary research interests are the delivery of municipal services to neighborhoods, neighborhood change and social indicators, the social service impact of telecommunications, and the relationships between citizens and governments. He holds a B.A. in history from Harvard, an M.A. in government and public administration from George Washington University, and a Ph.D. in social psychology from MIT.